MW00639346

THE TALIBAN'S VIRTUAL EMIRATE

The Taliban's Virtual Emirate

THE CULTURE AND PSYCHOLOGY OF AN ONLINE MILITANT COMMUNITY

Neil Krishan Aggarwal

COLUMBIA UNIVERSITY PRESS NEW YORK

COLUMBIA UNIVERSITY PRESS
Publishers Since 1893
New York Chichester, West Sussex

Library of Congress Cataloging-in-Publication Data
Names: Aggarwal, Neil Krishan, author.
Title: The Taliban's virtual emirate : the culture and psychology of an online militant community / Neil Krishan Aggarwal.
Description: New York : Columbia University Press, [2016] | Includes bibliographical references and index.
Identifiers: LCCN 2015035460| ISBN 9780231174268 (cloth : alk. paper) | ISBN 9780231541626 (electronic)
Subjects: LCSH: Virtual reality—Religious aspects—Islam. | Taliban. | Virtual reality—Political aspects—Afghanistan.
Classification: LCC BP190.5.V57 A34 2016 | DDC 302.23/109581—dc23
LC record available at http://lccn.loc.gov/2015035460

Columbia University Press books are printed on permanent and durable acid-free paper.
This book is printed on paper with recycled content.
Printed in the United States of America

c 10 9 8 7 6 5 4 3 2 1
p 10 9 8 7 6 5 4 3 2 1

Cover design: Kei Kato/Fifth-Letter
Cover image: iStockphoto

For purposes of our discussion here, cosmopolitan and vernacular can be taken as modes of literary (and intellectual, and political) communication directed to two different audiences whom lay actors know full well to be different. The one is unbounded and potentially infinite in extension; the other is practically finite and bounded by other finite audiences, with whom, through the very dynamic of vernacularization, relations of ever-increasing incommunication come into being. We can think of this most readily as a distinction in communicative capacity and concerns between a language that travels far and one that travels little.

—SHELDON POLLOCK, "Cosmopolitan and Vernacular in History," 2000

CONTENTS

ACKNOWLEDGMENTS *IX*

TRANSCRIPTION AND TRANSLATION GUIDE *XI*

PREFACE *XIII*

CHAPTER ONE

▸ Channels of Communication in the Virtual Emirate *1*

CHAPTER TWO

▸ Mullah Omar's Leadership in the Virtual Emirate *30*

CHAPTER THREE

▸ Identity in the Virtual Emirate *64*

CHAPTER FOUR

▸ Jihad in the Virtual Emirate *90*

CHAPTER FIVE

▸ International Relations in the Virtual Emirate *117*

▸ Epilogue *143*

NOTES *149*

REFERENCES *159*

INDEX *203*

ACKNOWLEDGMENTS

I AM GRATEFUL to those who have supported both me and this book.

Generous colleagues: Hussein Abdulsater, Muzaffar Alam, Ali Asani, Alex Barna, Adia Benton, Homi Bhabha, Kam Bhui, Byron Good, Mary Jo DelVecchio Good, Ezra Griffith, Hamada Hamid, Schuyler Henderson, John Horgan, James Jones, Emily Keram, Marilyn King, Arthur Kleinman, Orla Lynch, Matt Melvin-Koushki, Sarah Pinto, Jerrold Post, Annelle Primm, Dan Sheffield, Wheeler Thackston, and Stephen Xenakis. I have edited this book as Shahab Ahmed has succumbed to illness; I hope that he sees his influence in my work and that we joke in the Afterlife about him playing soccer with Osama bin Laden in Afghanistan in the 1990s.

Wonderful friends: Radhu and Meena Agrawal, Rohit Agrawal, Ayesha Ahmed, Iqbal and Sheela Bakhshi, Moustafa Banna, Omer Bokhari, Rachna Dave, Ravi DeSilva, Ashok and Veena Dhar, Lahari Goud, Yusuf Iqbal, Usha and Raj Machiraju, Samir Rao, Sapan Shah, Nabilah Siddiquee, Luvleen Sidhu, Satbir Singh, Vinita Srivastava, Rizwan Syed, and Parvinder Thiara. Also: the Hindu-Jain Temple and Gujarati Samaj of Pittsburgh, Pennsylvania, which have generously publicized my work.

Columbia University Press: Jennifer Perillo, Stephen Wesley, and all the extraordinary people who convert book proposals into books.

My supportive family: in-laws Niraj Nabh and Anshu Kumar, parents Madhu and Krishan Kumar Aggarwal, sisters-in-law Reema Aggarwal and Radhika Kumar, brother Manu Aggarwal, niece and nephew Asha and Roshan Aggarwal, daughter Amaya Ishvari Aggarwal, and wife Ritambhara Kumar.

This book is dedicated to three couples: my parents and my grandparents, Kesar Devi, Nanak, Sarla, and Shyam Lal Aggarwal. Like millions of refugees in post-Partition India and Pakistan, they struggled with new alignments in culture, language, and religious identity in turbulent political times.

TRANSCRIPTION AND TRANSLATION GUIDE

THIS STUDY EXAMINES the circulation of psychological and cultural themes across Taliban texts in four languages, namely English, Arabic, Dari (the Afghan dialect of Persian), and Urdu. I retain Arabic transcription for Arabic loan words in Dari and Urdu unless these words are pronounced differently. The latest *Chicago Manual of Style* (16th edition) indicates a growing scholarly trend toward removing diacritical marks from letters for simplification—a style adopted in this book.

Common words such as *Hadith* (the texts on the Prophet Muhammad's sayings and doings), *jihad* (literally "struggle," though often translated as "holy war"), *madrasa* (religious school), *mujahideen* (holy warrior), *Shari'a* (Islamic law), *Quran*, and *ummah* (the Muslim confessional community) appear in customary Anglicized form. *Allah* and *Khuda* are both translated as "God," and *Allāh ta'ālā* is translated as "God (may He be exalted)." Technical terms are transcribed only if rare or obscure, but full transcriptions are presented in the reference list. Names are only transcribed to differentiate vowel sounds—for example, Rāshid versus Rashīd. I keep the Taliban's own transcription for its Arabic periodical as *Al-Somood*, rather than *Al-Sumūd*.

I do not translate Quranic text, since Quranic exegesis is an ongoing body of scholarship rife with debate on textual interpretation. Instead, I cite chapter and verse, deferring to A. J. Arberry's standard *The Koran Interpreted: A Translation*.

PREFACE

IN MY PREVIOUS BOOK, *Mental Health in the War on Terror* (Aggarwal 2015), I explored the war's effects on mental health knowledge and practice and posited that culture-specific values concerning acceptable thoughts, emotions, and behaviors have influenced the practice of terrorism. In a chapter on suicide bombing, I juxtaposed representations of suicide bombers from mental health scholars with those of Al Qaeda writers to illustrate fundamental discrepancies in common human experiences such as life, death, illness, and suffering.

While the search for Al Qaeda texts took weeks of steadfast searching, as websites migrated, mandated new passwords, or disappeared altogether, Taliban websites defiantly and publicly proclaimed their perspectives. These multilingual websites are not just simple translations from one language into another; English authors attacking Western political philosophy may cite Robespierre and Noam Chomsky while Urdu authors will reference Muhammad Iqbal and Akbar Allahabadi. Taliban authors engage with the War on Terror on their own terms, fashioning texts with viewpoints that follow their respective cultural models for argumentation. For example, a couplet of canonical poetry meant to reinforce an author's theme might be considered trite in a scholarly English article but would be stylistically acceptable in Urdu writing. Because a thorough treatment of this phenomenon would have overwhelmed the first book, I realized that another volume would be necessary.

An analysis of the Taliban primarily through its own sources and in relation to secondary scholarship gives an alternative account of the War on

Terror that would be of interest to educated readers; regional scholars of South Asia and the Middle East; and topical specialists in cultural psychology, history, political science, and international relations. Unfortunately, we have little scholarship on Taliban texts compared to Al Qaeda texts (Ciovacco 2009; Payne 2009; Ryan 2013), despite their wide availability after fifteen years of war in Afghanistan. Like the Taliban, Al Qaeda (Loidolt 2011), Hamas (Mozes and Weimann 2010), and Chechen insurgents (Knysh 2012) have also published in multiple languages to reach global audiences (Tsfati and Weimann 2002).

The Internet has destabilized geographical notions of "local" and "global," as readers anywhere can access, identify, and attach themselves to distant conflicts, compelling us to rethink relationships among language, cultural identity, and psychology. Sheldon Pollock's quote in the epigraph confronts the sociopolitical ramifications of writing texts in languages that travel little compared to those that travel widely. Literary culture—reading, writing, and circulating texts—authorizes forms of attachment and affiliation, both of which are affective and cognitive operations. Taliban authors try to elicit these affective and cognitive responses by invoking cultural tropes. In this book, I analyze multilingual texts created by the Taliban to uncover how it markets its messages. This examination attempts to disentangle and reformulate the relationships among language, cultural identity, and psychology by thinking through the role of militant literature in promoting group formation of an ideology.

As of yet, I have not found a satisfactory framework for gathering and analyzing texts for an interdisciplinary project in the cultural and social sciences. Consequently, I have conducted research through my own series of steps, each with reproducible and generalizable methods:

Step 1: Finding Taliban websites. Needless to say, the Internet contains unreliable information. Further, search engines may retrieve irrelevant results, especially from militant groups whose Internet websites may not be easily authenticated. Data triangulation can solve these problems. Precise key terms should be entered into search engines, websites should be checked against secondary scholarship when possible, and back-link search functions within search engines should point to source websites (Chen et al. 2008).

Step 2: Locating the texts. I analyze texts in a variety of formats posted on the Taliban's websites and available for download, such as articles, songs, and videos. These texts often claim to represent official viewpoints. The Urdu monthly *Sharīʿat*, for example, declares itself "the sole mouthpiece" (*wāḥid tarjumān*) of the Taliban. I primarily analyze speeches from the late supreme leader Mullah Omar; official statements; scriptural exegeses; interviews with officials; and articles of social, political, and cultural interest. I treat Taliban texts as an archive of state documents from its self-styled Islamic Emirate of Afghanistan, in line with textual methods in Islamic studies (Humphreys 1991). That no state recognizes the emirate is irrelevant; the unique historical situation of a militant organization coming to power, falling, and then maintaining a parallel ghost government heightens my curiosity about what Taliban authors seek to accomplish through their texts.

To test the hypothesis that Taliban authors write for readers who differ linguistically and culturally, I analyze periodicals by language: all 104 issues of the Arabic *Al-Somood* since its start in 2007, all 17 issues of the bilingual Dari-Pashto *Srak* since 2011, all 5 issues of the Dari *Haqīqat* since its start in 2014, all 6 issues of the English *Azan* since its start in 2013, and all 34 issues of the Urdu *Sharīʿat* since its start in 2012. These are not all the Taliban texts in each language, but they represent primary sources that have never been systematically explored.

Step 3: Rethinking the relationship of language to cultural identity in multilingual communities. Some claim that the Internet's global reach has antiquated the relationships among language, territory, and cultural identity (Urciuoli 1995). However, the assumption that underlies this claim—namely that language has at any time been a singular means of grounding identity—originated in nineteenth-century European sociolinguistic scholarship, influenced by the extreme European nationalism of the time (Gumperz and Cook-Gumperz 2008). In contrast, South Asians have been fluent in multiple languages for centuries, and language alone has never determined cultural identity (Pollock 2006). Instead, scrutinizing the circulation of textual themes in separate linguistic audiences can elucidate aspects of the Taliban's transnational constructions of cultural identity more clearly (Orsini 2012). I position my book in this tradition. Some may object that Afghanistan is part of Central Asia, not South Asia, but I consider Afghanistan to be part of South Asia, following historians (Wolpert 1982; Bose and Jalal 2004), American "Af-Pak" foreign policy (White House 2009; I. Ahmad 2010), and Afghan officials themselves (South Asian Association for Regional Cooperation 2014).

Step 4: Selecting appropriate texts for analysis. My training in both cultural psychiatry—a medical specialty centered on communication—and

South Asian studies shapes my view that language and discourse present valuable avenues to analyze identity (Aggarwal 2011). The importance accorded to language assumes that all humans can share thoughts intelligible to one another (Sinha 2000) and that language is the dominant means through which members of society share meanings (Wan 2012). Language is necessary but not sufficient to map the contours of cultural identity without triangulating data on language use in society—such as literary themes and tropes—through discourse analysis. Discourse analysis has illustrated the use of language to promote ideology and influence public opinion through newspapers, speeches, and interviews (Perrin 2005; Gamson and Herzog 1999; Gray and Durrheim 2013; Weltman and Billig 2001; De Castella, McGarty, and Musgrove 2009). I assume that Taliban texts are assemblages of discourse promoting specific ideologies to influence public opinion through cultural meanings, understandings, and expectations.

The number of Taliban periodicals raises questions about which texts to analyze. Each Urdu periodical has about fifty-two pages, and there have been thirty-four issues as of January 2012, leading to over 1,768 pages for this language. To isolate relevant texts, I followed a classification for militant Islamist organizations based on five domains: (1) channels of communication, (2) leadership hierarchy, (3) identity of members, (4) organizational ideology, and (5) targeted enemies (Mishal and Rosenthal 2005). These domains yield a skeletal outline for each chapter but must be fleshed out for unique properties with respect to the Taliban. I know of no method or framework that connects these domains, so I have brought this classification into dialogue with theories from cultural psychology and psychiatry, Arabic and Islamic studies, and South Asian studies. I have screened the titles and first paragraphs of all articles for inclusion and exclusion by these domains—a method used previously in similar studies (Aggarwal 2010, 2015).

Chapter 1 introduces readers to the Taliban in the way that the Taliban has announced its presence to the world: through Internet-based channels of communication. Here, I also engage with the extensive scholarship on the growth and evolution of the Taliban. Right from the outset, the Taliban has promoted Mullah Omar as the most virtuous Muslim and the undisputed leader of the organization. Hence, chapter 2 uses the Taliban's focus on Mullah Omar to analyze the psychology of leadership in the organization. This chapter combines methods in political psychology to analyze leadership through the Taliban's own texts. Official statements attributed to Mullah Omar on who counts as an insider, the mission of the organization, and enemies to be targeted are the bases for chapters 3, 4, and 5, respectively. There is considerably less scholarship in these areas, so readers will see that I draw more extensively on the Taliban's writers in these chapters to formulate and test hypotheses.

Because there are hundreds of texts to review, how many analyses would be needed to ensure an accurate representation of the Taliban's culture? For an answer, I turned to social scientists who study data saturation: the extent to which we know when we have accurately described a phenomenon. Most agree that analyses of six to twelve examples can be considered an accurate representation of themes that circulate within a culture, provided that (1) texts come from a homogeneous group of people, (2) texts are produced independently, and (3) research criteria form a coherent field of knowledge (Guest, Bunce, and Johnson 2006; Onwuegbuzie and Collins 2007). I make the following assumptions: that the Taliban acts as a homogeneous group, that its authors write independently, and that this classification of militant Islamist organizations is coherent (Mishal and Rosenthal 2005), since all militant organizations have leaders, followers, a mission, an enemy, and a means of communication.

I have analyzed at least six texts for every example in each chapter (unless otherwise stated) and have included more when authors represent themes differently. I reproduce quotes to let Taliban authors speak for themselves, drawing on secondary scholarship on Afghanistan to introduce chapters and situate Taliban references. Readers looking for more detailed histories of Afghanistan can peruse the many significant texts in the reference list. I restrict my focus to the Afghan Taliban, since Pakistani Taliban groups are more heterogeneous, with varying leaders, followers, organizational missions, and means of communication.

Step 5: Translating the texts. I have translated all texts from Arabic, Dari, and Urdu unless otherwise noted. Translators convert texts differently based on purpose, sometimes striving for strict formal equivalence of a message's form and content, sometimes preferring dynamic equivalence for equal impact on the audience. Effective translations should (1) make sense, (2) convey the spirit of the original text, (3) have an easy and natural form of expression, and (4) produce a similar response for readers (Nida 2000). Translators should also consider the politics of representing foreign communities and be sensitive to power differences (Spivak 2000). The act of translation attempts to render foreignness in native terms, but there will always be elements that defy translation (Bhabha 1994). I view "foreignness" not as a gulf but as a bridge toward comprehension. When difficulties in translation arise, I transcribe words from original languages with explanation.

Step 6: Analyzing the texts. Cultural psychiatrists have encouraged research on how the discourse of militant groups transmits cultural meanings and persuades potential recruits (Bhui and Ibrahim 2013; Kirmayer, Raikhel, and Rahimi 2013). I accept Michel Foucault's (1991) points that (1) those in power produce knowledge with claims to authority, (2) this knowledge reinforces positions of power, and (3) the type of knowledge that trumps all others is an official discourse. Studying discourse through texts can define the boundaries of the subjects' knowledge, reveal how language acquires new function, and track the circulation of ideas in society (Foucault 1991, 56–57). Critical discourse analysis allows us to contemplate how social and cultural theories extend to new circumstances by assuming that some members in a culture dominate others by controlling communication (Fairclough 1992, 2001; Van Dijk 1993). I assume that texts comment on social actions, representations of the self, and representations of others, with meanings of texts reflecting the world outside of texts (Fairclough 2001). A representation is any idea understood through language (Potter and Edwards 1999), and it is social if it is shared in two or more minds (Farr 1998). I assume that Taliban texts are designed for circulation and represent the world outside of texts as perceived by the authors.

Discourse analysis treats culture as a psychological phenomenon through shared symbols, concepts, and meanings. This view of culture excludes theories of culture organized nonlinguistically—for example, through observable behaviors, material living conditions, and social institutions (Ratner 1999, 2008). These aspects of culture would be better analyzed through ethnography and participant observation. I have also concentrated on texts to avoid legal consequences of interviewing enemies of America (Savage 2013a, 2013b). My analyses are subjective, since researchers can analyze the same texts through different theories. This is both a strength pointing to creativity and a weakness pointing to a discipline-specific idiosyncrasy in the interpretive social sciences (Rabinow and Sullivan 1979). There are practical limits to the number of texts I can analyze; therefore, my conclusions should be regarded as representative rather than exhaustive. The materials surveyed here run to thousands of pages, and if past trends hold true, the Taliban will publish texts after this manuscript is published. I have excluded many fascinating texts for not meeting strict inclusion and exclusion criteria.

Step 7: Limiting biases throughout analyses. Research that involves group-based categories such as race, culture, ideology, or nationality may be biased if the categories are only relevant for the researchers and not the communities studied (Gillespie, Howarth, and Cornish 2012). To avoid this problem, I focus on the goals, values, and depictions of the world transmitted through texts to understand a community's way of life (Shweder 1999). Granting interpretive charity requires a willingness to suspend what we believe to be true to follow how others construct their worlds (Shweder 1991, 1997). It has not been easy for me to grant the Taliban the benefit of the doubt, since it has targeted several of my group-based categories, including *American*, *Hindu*, and *Indian*. We do not come to research devoid of preexisting feelings or theories. It would be impossible to negate years as an Indian American Hindu living in a society dominated by the rhetoric of the War on Terror or as an academic writing within a mental health tradition that often equates militants with terrorists without first soliciting their perspectives (Aggarwal 2010). Here I attempt to avoid imposing my agenda on the Taliban. Ultimately, readers will judge my success.

THE TALIBAN'S VIRTUAL EMIRATE

CHAPTER 1

Channels of Communication in the Virtual Emirate

BEFORE THE ISLAMIC STATE publicized the immolation of its enemies on the Internet in 2015, and before Al Qaeda in Iraq disseminated videos of the decapitation of Americans online in 2004, the Taliban launched its website in 1998 (www.taliban.com). At that time, I was conducting research for an undergraduate thesis on militant movements in postcolonial South Asia. Aside from the Taliban, the only militant group to host a website was the Liberation Tigers of Tamil Eelam in Sri Lanka. The Tigers used a basic point-and-click format, but the Taliban featured cutting-edge technologies; its home page opened with digitized American, Israeli, and Indian flags zooming into central view, exploding, then receding into the background to be overlaid by scrolling text in English and Urdu. Seventeen years later, the Tigers have been tamed by the Sri Lankan military, while the Taliban's website for the Islamic Emirate of Afghanistan (www.shahamat.com) now includes text and multimedia sections of audio and video content in English, Arabic, Dari, Pashto, and Urdu. The multimedia sections indicate that the Taliban has been able to invest even more resources than before to host large files, indicating its ongoing popularity. Unlike other South Asian militants, the Taliban has expanded ambitiously to reach an audience in more languages than ever before. The name of the website is also illustrative: Arabist Hans Wehr defines the word *emirate* as "principality, emirate; authority, power" (1976, 27). The Taliban's name for its website signals its ambitions for Afghanistan, envisioning a type of sovereignty based on an Islamic political system.

The Taliban clearly takes pride in its websites. In an interview posted on the Taliban's official English website, Moulavi Mohammad Saleem—deputy head of the Cultural Commission for the Islamic Emirate of Afghanistan—boasted that the Taliban's multilingual websites are the most read in Afghanistan. Here, I reproduce the original quote, which contains grammatical errors:

> If we take the internet department, the Islamic Emirate has a total of ten sites pulled together into one site. Pashto, Dari, Arabic, English, Urdu, Islam, video site, anthems, Al-Samood and Magazines site; all these sites are refreshed on the daily basis and fresh material is uploaded to them. Approximately more than 50 news items from all over Afghanistan is published in the sites of five languages i.e. Pashto, Dari, Arabic, Urdu and English. In addition, fresh incidents of the country and the world, interviews, reports, political and analytical articles, weekly analysis, messages and other material of literary touch is published. If we evaluate by the volume of the daily fresh news and moment to moment edit, we can say with full satisfaction that Alemarah ["the Emirate" in Arabic] site is the richest and widest read site on the level of Afghanistan. (Islamic Emirate of Afghanistan 2013a)

Table 1.1 lists the total number of hits for the most accessed web content by language since publication on the Taliban's websites. The Urdu news section article boasts close to seven thousand hits, compared to nearly nine hundred for the English news section article. The most-accessed statement in English exceeds ten thousand hits, more than that for the Arabic, Dari, and Urdu statements combined. These variations suggest that there are differences in how people access this content, with Urdu speakers possibly relying on the Taliban's websites for primary reporting compared to speakers of other languages. The high number of hits for the English statement may be due to the number of readers abroad, such as supporters, journalists, scholars, and people who are curious about the Taliban's Internet presence. Reports of military triumphs in Afghanistan coexist with statements on international affairs (i.e., Egypt), suicide bombing, and Islamic finance. This diverse range of topics points to the role of the Taliban's websites in promoting a distinctive worldview beyond the war in Afghanistan.

Furthermore, if we take the Taliban at its word, its websites front an enormous enterprise. This interview fragment implies an expensive and expansive undertaking. This quote is longer than others in the book, but

TABLE 1.1 Content with Most Accessed Hits by Language on the Taliban's Websites*

	ENGLISH	ARABIC	DARI	URDU
NEWS TITLE	Massive Blasts Hit US Installations in Jalalabad Airbase (859 hits)	The Latest: Destruction of 3 Planes and the Targeting and Killing of a Large Number of American Soldiers in a Bombing of Bagram Airbase (2,771 hits)	The Tragic Ongoing Revenge of Siagard (1,680 hits)	The Video of the Prisoner Exchange Between the Islamic Emirate and American Government (6,920 hits)
WEEKLY ANALYSIS	Karzai in the Vortex of Frustration (4,324 hits)	The Hiring of the Republic's Leadership, and the Important Problem of the People (4,515 hits)	The Islamic Emirate and Standards for Ties with Countries (2,020 hits)	Presidential Candidates and an Important National Problem?!! (3,738 hits)
ARTICLES	Let's Understand Suicide Bombing (6,210 hits)	Fleeing of the Enemy from Farah Province, Report from Correspondents of *Al-Somood* (3,945 hits)	Islamic Economics and Islamic Banking (2,058 hits)	"Shattering False Gods": The 36th Film Release of The Emirate's Jihadist Studio (5,503 hits)
INTERVIEWS	Symptoms of Great Change in the Shi'ite Populated Areas (6,878 hits)	Interviews with the Official Responsible for General *Mujahideen* in Pakitya Province (7,497 hits)	A Succinct Look at the Line of Education in the Province of Wardak (2,123 hits)	268 Officials Surrender in January 2014 (4,133 hits)
STATEMENTS	The Islamic Emirate of Afghanistan Financial Commission (10,226 hits)	Statement of the Islamic Emirate of Afghanistan on the Continuing Massacres in Egypt (3,795 hits)	Statement of the Leadership Council of the Islamic Emirate on the Topic of Starting the Spring Operations "Khalid bin Walid" (3,095 hits)	Clarification of the Spokesperson on the Cutting of Fingers of Some People in Herat (820 hits)

* Statistics were tallied on June 27, 2014. The Taliban has not listed statistics for Pashto websites.

I have found no other text that better encapsulates the range of activities or resources of the Taliban and gives us a sense of its day-to-day activities. Even if this text is pure propaganda, it imparts an understanding of the types of values that the Taliban prioritizes for its audiences:

> At present, the high commissions of the Islamic Emirate are as following though all of them have vast set-up[s] within themselves: Courts, Military Commission, Commission of Cultural Affairs, Education Commission, Invitation and Guidance Commission, Health Commission, Organ for the Prevention of Civilian Casualties, Commission for Prisoners' Affairs, Organ for Martyrs, Orphans and Disabled persons, Commission for Political Affairs and Commission for NGO's Affairs.
>
> As far as the responsibilities are activities of these commissions are concerned, I would like to give you the example of [the] "Health Commission." This commission is responsible for [the] wounded, paralyzed, [and] disabled as well as the treatment of those Mujahidin who are released from jails. You can imagine that the number of wounded and disabled people in the prolonged battle of Afghanistan is quite large[,] and every year it is increased by thousands. According to the estimations of the "Health Commission," the average expenditure of every wounded, disabled or paralyzed one reaches to 200,000 Afghanis which is equal to more than 4000 dollars. The expenses of some seriously wounded persons soar several times higher than this.
>
> The "Commission for Prisoners' Affairs" provides cash money to the incarcerated people in all jails of Afghanistan. You might be aware that tens of thousands of innocent Muslims are lying in various jails of Afghanistan without any charges. Being a responsible and committed Islamic movement, the Islamic Emirate of Afghanistan reckons it one of its foremost responsibilities to assist all the incarcerated people within the range of its possibilities. Therefore, it has established a separate organ for this purpose to support the oppressed detainees financially for getting treatment and meeting daily expenses.
>
> The "Commission for Martyrs, Orphans and Crippled Affairs" is responsible for collecting the lists of all those children who are orphaned after the American invasion in all the 34 provinces of Afghanistan. Their detailed bio-data is assembled; then, they are supported. Similarly those people, who are paralyzed or disabled undyingly, are also supported finan-

> cially on the permanent basis. There is [a] large number of people in our society who lost both of their legs, hands or both eyes. They do not have any source of income apart from the Islamic Emirate to support their families and offspring.
>
> Besides supporting them financially, the "Educational Commission" of the Islamic Emirate has also set up orphanage centers for them so that these helpless children should not remain uneducated. Similarly, the "Invitation and Guidance" or "Attraction and Absorption Commission" is responsible for guiding those people who are working side by side with the American infidels. This commission tries to separate these people from the enemy. The "Commission for the [P]revention of [C]ivilian [C]asualties" tries its best to save civilian and defenseless Afghans from war miseries. In case the civilian losses occur, this commission is responsible to financially assist the suffering families. Likewise, all the above[-]mentioned commissions perform their assigned duties and have heavy expenditure. (Islamic Emirate of Afghanistan 2014c)

In order to compete with the Afghan government and be seen as a viable political alternative, the Taliban proudly advertises its expansive administration, flush resources, and vast network. This undertaking finds a historical precedent in the mujahideen resistance movements of the 1980s, which created alternative systems of government to the Soviet Union–backed Kabul regime in approximately 80 percent of the country (Amin 1984). This quote also illustrates the Taliban's ideological mission to rule over Afghanistan once again, as it did from 1996 to 2001. These types of texts from Taliban authors can teach us about their military strategy, sociopolitical perspectives, and campaigns for recruitment. Currently, we have little academic research on Taliban texts, besides cursory overviews of Taliban periodicals in different languages (Chroust 2000; Nathan 2009). This trend reflects the general state of scholarship on the Taliban, with most studies relying on American military and civilian sources (Farrell and Giustozzi 2013). Instead, studies exploring the Taliban's internal perspectives have typically involved interviews with Taliban leaders to explain the group's post-2001 resurgence (Giustozzi 2009, 2012; Malkasian 2013; Strick van Linschoten and Kuehn 2012a; Bergen and Tiedemann 2013). The few textual studies include translations and analyses of a former minister's autobiography (Zaeef 2010), original poetry from rank-and-file members (Strick van Linschoten and Kuehn 2012b), and the *Layeha*, a military code of conduct for militants (Shah 2012; Johnson and DuPee 2012). We know little about the Taliban worldview, even though the wars in Afghanistan and Iraq have surpassed $1.6 trillion and 145,000 deaths (Watson Institute 2013), and American military involvement in Afghanistan has exceeded all other wars. For example, while we know that the Taliban has fought American-led coalition troops as enemies, we do not know how they have managed to stage a successful resistance over the years. Analyzing Taliban texts allows us to observe cultural and psychological trends of the organization as they relate to a wide range of issues.

Surprisingly, the Taliban's multilingual publications demonstrate a trend of ideological flexibility that sharply deviates from its history of cultural, social, and political rigidity. In the next section, I synthesize Taliban primary sources and key works of secondary scholarship to chart the Taliban's trajectory from the 1990s to contemporary times. This information provides a context in which to consider why a militant organization pledging to replicate the conditions of seventh-century Arabia in Afghanistan has sought to exploit twenty-first-century multimedia technologies so effectively.

THE GROWTH AND EVOLUTION OF THE TALIBAN

After the 1989 retreat of the Soviet Union from Afghanistan and the 1992 fall of the Soviet-backed Mohammed Najibullah government, Islamist resistance groups—known as the mujahideen ("warriors" in Arabic) and previously funded by the United States, Pakistan, Iran, and Saudi Arabia—turned against one another to carve the country into exclusive spheres of influence (Nojumi 2008). The Taliban formed in 1994 to oppose these groups, which had taken to plundering populations and raping minors, vowing to disarm rival commanders, establish peace, and purify society by implementing Islamic Shari'a law throughout Afghanistan (Marsden 1998; Goodson 2012). Taliban radio broadcasts reiterate the centrality of jihad to justify violence, whether against atheist Soviet Communists or enemy warlords accused of conspiring with foreign governments (Kleiner 2000). The Taliban's website in 1998 portrayed a scenario perceived as wanton depravity:

> Conditions deteriorated to such an extent that the people cried out in anguish. They were now [ready] even for Russia or a power even much worse

> to take command of their country. They only asked to be delivered from their sufferings. Their misery know [*sic*] [no] bounds. No one's wealth, life or honour was safe anymore.
>
> Cities became deserted after Asr prayers, before the sun had even set. People did not dare to venture outside their homes. Tired after the day's work, they had yet to keep awake to guard their homes. Even then they were not safe. During the daytime too no one dared to venture out with their wives, daughters or sisters, out of fear for their honour and safety.
>
> People had sent their handsome, teenage sons to Pakistan as it had become extremely dangerous for them to venture out in the streets. Homosexuality, adultery had become so common that it is said a boy married another boy in Kandahar. Many people joyously celebrated the wedding, ringing the death-knell of all decency, all that is good and honourable. (Taliban 1998)

The Taliban represents the time before its rule as a dark period of desperation, deserted cities, dishonor, and decadence, with constant insecurity. Immorality is described as sexual relationships outside of any heterosexual marriage, whether in the form of adultery or homosexuality. No division exists between public and private lives in this commentary: all forms of personal behavior that are deemed immoral are also connected to crimes committed in public.

The Taliban soon attracted international attention. American officials believe that the Taliban's initial funding came from Afghan traders who wanted to smuggle goods from Central Asia to South Asia through Afghanistan. The civilian Pakistani government of Benazir Bhutto later provided jets, diesel fuel, ammunition, and artillery and weapons training (Holzman 1995). Other Pakistani military officials increased their support throughout 1994 by using private transportation companies to supply "munitions, petroleum, oil, lubricant, and food" (U.S. Intelligence Information Report 1996). Many Pakistani civilian and military elites commonly viewed the Taliban as a force to purify Islam of heretical groups through jihad (Cohen 2004). The Taliban recruited hundreds of Pashtun students from madrasas (Islamic schools) in Pakistani refugee camps. By 1998, these students had imposed Shari'a law, modeled on the Prophet Muhammad's society of seventh-century Arabia, on 90 percent of Afghanistan (Rashid 2002). Pakistan's Grand Mufti Rasheed Ahmad Ludhyanvi praised the

Taliban's attempt to implement Shari'a law: "In the areas under the Taliban Government every kind of wickedness and immorality, cruelty, murder, robbery, songs & music, TV, VCR, satellite dish, immodesty (*be-purdagi*) [literally 'being unveiled'], travelling without a *mehrum* [a male companion for women], shaving-off or trimming the beard, pictures & photographs, interest, have all been totally banned" (Moosa 1998). Most of the Taliban's leadership passed through two Pakistani madrasas, the Darul Uloom Haqqani at Akora Khattak and the Jamiatul Uloomil Islamiyyah in Karachi (Matinuddin 1999).[1] Taliban writers have proudly proclaimed these affiliations:

> The Taliban leadership, Al hamdu-lilLah [Praise God] comprises of Saheeh-ul-Aqeedah [True of Faith] Deobandi Muslims. They proclaim themselves that they belong to the Maslak-e-Deoband [Path of Deoband] which admits to no excesses but is firmly based upon moderation. We ourselves are a witness to the fact that the common Taliban leaders are former students of famous Deobandi Madaris [schools] of Pakistan. Jami'a Haqqaniyyah, Akhora Khattak, Peshawar; Jami'at 'Uloom-ul-Islamiyyah, Binnori Town; Dar-ul-Uloom Karachi; Jamia Farooqiyyah; Jamia Hammadiyyah; Dar-ul-Ifta-e-Wal Irshad, Nazimabad and other Madaris of Pakistan and Afghanistan have been their seats of learning. (Islamic Emirate of Afghanistan 1998a)

The references to "Deoband" and "Deobandi" refer to the educational orientation of the madrasa network attended by the Taliban. Begun in 1867 in Deoband, India, the Dār ul-'Ulūm seminary adopted a standard syllabus of Quran and Hadith study, central exams, and professional staff in all branches, promulgating a common theology throughout Muslim South Asia that contrasted with an older model of oral learning between teachers and students (Metcalf 1978). Since 1914, Deobandi scholars have triggered political mobilization in Pakistan's Pashtun territories by controlling mosques and madrasas. Abdul Haq, founder of the aforementioned mosque at Akora Khattak, oversaw the graduation of 370 Afghan students between 1947 and 1978 and converted his mosque into a center of jihad after the Soviet invasion (Haroon 2008). Despite the Taliban's claims of adhering strictly to Deobandi principles, leading Deobandi scholars in Pakistan have openly disagreed with the Taliban, encouraging it to educate girls and to resolve the crisis with the United States over Osama bin Laden after 9/11 in the best interests of Pakistan and Afghanistan (Zaman 2002).

The mention of Pakistan in the Taliban's quote also reflects disruptions in traditional Afghan society. Power dynamics have traditionally divided urban political elites who represent the state and rural tribal chiefs and spiritual leaders (Ghani 1978; Roy 1990). Religious scholars worked as state employees without an independent followership until the 1960s, when they demanded political participation—inspired by Islamist movements in Egypt, Pakistan, and Iran—and conducted political activities underground in the 1970s due to opposition from the Soviet-backed Kabul regime (Naby 1988). The erosion of social institutions during the 1980s war—such as the landlord-dominated rural economy and the state madrasa network—decentralized education to local mosques, whose leaders looked to religious scholars now housed in Pakistan (B. Rubin 2002). Many madrasa students in Pakistan were Afghan children who lost parents during the war; thus, the Taliban can be seen as an orphan youth movement transcending ties of tribe and kinship to form a group identity based on religion (Fergusson 2010).

In addition to seeking official government support, the Taliban has appealed to Pakistani civilians. The following 1998 appeal from the Al-Rasheed Trust calls for donations of animals for helping hungry and destitute Afghans.[2] The trust's branches in major Pakistani cities attests to the Taliban's national nexus:

> Al-Rasheed Trust requests all Muslims in general, supporters of the Taliban and readers of Dha'rb-i-M'umin [a periodical] in particular, to offer their "Wajib" [obligation] and "Nafl" Qurbani [sacrifice], in Kabul, Herat, Badghis and Kandhar [*sic*], on the occasion of 'Eid-ul-Adhha. Besides the "Wajib" of Qurbani, they would thus be discharging their "Fardh" [duty] of helping the poor and needy, the Taliban Mujahideen, families of the Shuhada [martyrs], thousands of destitute widows and orphans too. According to Mulla 'Abdul Jaleel Sahib, the cost of a medium sheep is Pak. Rs. 3,000/ or U.K. Pounds 40/ or U.S. Dollars 65/, only. The cost of the Qurbani can be deposited in the Trust's branches in Lahore, Rawalpindi, Peshawar, Mansehra, Mingora and its central office in Karachi. (Islamic Emirate of Afghanistan 1998b)

As a predominantly Pashtun ethnic movement, the Taliban has also shrewdly navigated the complex terrain of domestic identity politics. Afghanistan's predominant ethnicities include the mostly Sunni Pashtuns, who define themselves through a common lineage, moral code (known as *Pashtunwali*), and fluency in Pashto; the Sunni Tajiks, Shia Hazaras, and Sunni Aimaqs, who speak Dari; and the Sunni Uzbeks and Turkmens, who speak Turkic languages (Barfield 2010). These myriad groups contain internal kinship and regional rivalries, one reason why Afghanistan did not Balkanize during its turbulent years of war (Barfield 2010; Canfield 2011). Many ethnic boundaries appear relatively recent. Pashtuns have only rallied around the Pashto language as a marker of identity in the twentieth century, compared to previous Pashtun rulers who defined themselves by tribe and cultivated a Dari-speaking bureaucracy (Schiffman and Spooner 2012). In other words, Pashtun identity has gradually shifted from being ethnically based to being linguistically based. The term *Tajik* can variously refer to any Sunni Muslim, Dari speaker, or ethnic Tajik between the regions of Badakhshan and Faryab (Roy 2001).

Table 1.2 presents the percentage of ethnic groups in Afghanistan.

Scholars broadly agree that the Taliban fanned domestic Pashtun nationalism with support from certain elements within Pakistan, though they differ over why. Some believe that the Taliban stoked Pashtun triumphalism to counteract Tajik, Uzbek, and Hazara mujahideen groups after anarchy erupted in 1992 (Hyman 2002). The Taliban's interpretation of Islam also created a Pashtun identity through religion rather than tribe

TABLE 1.2 Ethnic Groups in Afghanistan

ETHNIC GROUP	PERCENTAGE
Pashtun	42%
Tajik	27%
Hazara	9%
Uzbek	9%
Aimaq	4%
Turkmen	3%
Baloch	2%
Other	4%

Source: CIA World Factbook (2013). Afghanistan's population is estimated at 31,108,077.

and sidelined secular nationalism (Bijleveld 2008; Mishali-Ram 2011; Saigol 2012). Others dispute claims of pan-tribal Pashtun unity, since Taliban cadres came mostly from the Durrani tribe compared to the Ghilzai tribe–dominant, rival mujahideen movement of Gulbuddin Hekmatyar (Liebl 2007). Another argument against the Taliban as a purely Pashtun movement comes from the minutes of one confidential meeting between Taliban and American embassy officials in Pakistan, noting that the Taliban tried to recruit Tajiks (U.S. Embassy [Islamabad] 1996). Nonetheless, Pakistani government officials sought to protect their interests by counteracting Pashtun secessionists along the Durand Line border separating the two countries (Schofield 2011),[3] building a Turkmenistan-Afghanistan-Pakistan oil-and-gas pipeline (Shahrani 2002), and encouraging the dependence of Afghanistan's leadership on Pakistan's rulers (Weinbaum and Harder 2008). Facilitating these ties were Pashtuns in Pakistan's Inter-Services Intelligence (ISI) agency, who crossed the border to fight alongside the Taliban (Tomsen 2011), and Pakistan's Foreign Ministry, which provided six million dollars in salaries to senior Taliban officials (U.S. Embassy [Islamabad] 1998c). Figure 1.1 is a map of areas in Afghanistan under the Taliban's control before 2001.

The multilingual nature of the Taliban websites contrasts with policies favoring Pashtun language and culture during its brief time in power from 1996 to 2001. Dari and Pashto have been Afghanistan's official languages, with Dari a common medium among the country's ethnicities (Ahmadi 2008). During the rule of Abdur Rahman Khan (1880–1901), Dari became the de facto bureaucratic language adopted by Pashtun tribes for educational and economic advancement (Kakar 1995). Although many Pashtuns have learned Dari, many non-Pashtuns have not typically learned Pashto (Magnus and Naby 2002). In the 1920s, King Amanullah (1892–1960) promoted Pashto to galvanize support among Pashtun tribes along both sides of the Durand Line, whose border he disputed (Hussain 2005). In 1936, Muhammad Hashim, the younger brother of the late king Muhammad Nadir Shah (1883–1933), ordered that Pashto should replace Dari in all education (Saikal 2012). In the 1940s and 1950s, policies enforcing Pashto as the sole school language and mandating Pashto classes for non-Pashtun bureaucrats bred resentment among minorities (Entezar 2008). After the 1979 Soviet invasion, the Communist-backed government passed the "nationalities" policy to recognize Uzbek, Turkoman, Baluchi, and Nuristani

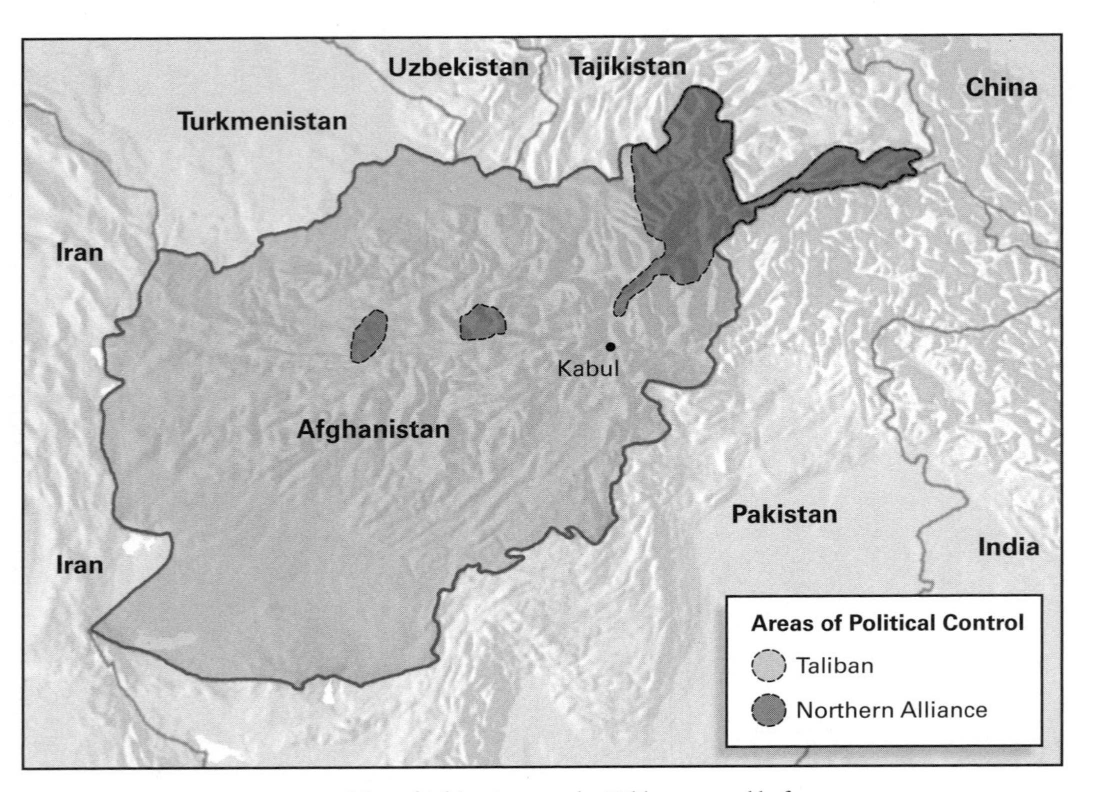

FIGURE 1.1 Map of Afghanistan under Taliban control before 2001.

as state languages, along with Pashto and Dari (Rasuly-Paleczek 2001; Rasanayagam 2007). The nationalities policy was intended to organize minorities against Pashtun mujahideen groups at the forefront of military resistance (Banuazizi and Weiner 1986).

The withdrawal of the Soviet Union in 1989 led to Afghanistan's geographic and ethnic division: Pashtuns dominated southern Afghanistan under Gulbuddin Hekmatyar, Uzbeks led by Abdul Rashid Dostum controlled the north, Tajiks under Ahmed Shah Massoud and Burhanuddin Rabbani held the northeast, and Afghan Shias fought under Ismail Khan in the south (Akcali 1998). Foreign powers protected their interests: Uzbekistan backed Uzbeks, Russia and Tajikistan backed Tajiks, and Iran backed Shia coreligionists (Malik 1999). The Taliban viewed Dari as a dialect of Persian and, consequently, representative of a Shia culture deemed heretical, emanating from its enemy Iran (Gohari 1999). Upon proclaiming sovereignty in 1997, the Taliban replaced non-Pashtun bureaucrats with Pashtuns, most who did not know Dari or the language of local constituents (Wahab and Youngerman 2007). For the first time in Afghan history, Pashto became the sole language of bureaucracy over Dari (S. Nawid 2012). As refugees in Pakistan, Taliban members learned Urdu as a second language rather than Dari, whose instruction was lost after the collapse of the Afghan educational system (Edwards 2002). In this light, the Taliban's substantial investment to maintain websites in English, Arabic, Dari, and Urdu differs pointedly from pro-Pashto policies during its short rule. The next section considers reasons for the Taliban's linguistic concessions after its overthrow in the post-9/11 period.

THE TALIBAN AFTER THE SEPTEMBER 11 ATTACKS

The Taliban's unswerving patronage of Pashtun culture and Sunni law before the 9/11 attacks contrasts with its subsequent capacity for adaptation. On September 11, 2001, news broke that the Taliban had assassinated Massoud—leader of the rival Northern Alliance, composed of all major non-Pashtun mujahideen and backed by Iran, Russia, India, and Tajikistan—and would mobilize twenty-five thousand troops for a final onslaught (Rashid 1999, 2001a). After the Taliban refused to relinquish Osama bin Laden, the United States invaded Afghanistan on October 7, 2001, to retaliate against Al Qaeda's 9/11 attacks and back the Northern

Alliance.[4] By 2004, Hamid Karzai's alliance with corrupt militias, civilian mistreatment from North Atlantic Treaty Organization (NATO) forces, and Pakistan's ongoing support for militancy led to the emergence of a "neo-Taliban," which relaxed strict adherence to religious commitments and initiated a new insurgency (Giustozzi 2008). The neo-Taliban unites two groups wishing to expel foreigners and establish an Islamic state in Afghanistan: those aligned with the transnational Islamist movement of Al Qaeda and those committed to Pashtun nationalism, believing the Karzai government to favor ethnic minorities (Tarzi 2009). High levels of unemployment, good pay for soldiers, and noninterference in poppy cultivation—a lucrative revenue source for rural farmers—have also contributed to neo-Taliban recruitment (Roberts 2009).

The neo-Taliban has used new tactics: it has retreated to Pakistani cities after attacks; purchased modern military equipment, such as durable boots and North Face jackets (rather than relying on slippers and shawls); produced close to eight thousand improvised explosive devices annually; encouraged rural education and vaccinations; and issued a code of conduct to local fighters (Gutman 2010). It has erected rural power structures based on opium production,[5] the enforcement of Islamic law, and militancy (Thruelsen 2010). It has also created shadow governments: a provincial governor oversees an Islamic judiciary of one judge and two scholars, along with a provincial commission, district mayor, and deputy district mayor under central command of the Quetta Shura (council), headed by the Taliban's supreme leader (Johnson 2013). Complicating matters, the Pakistani Taliban—Tehrīk-e-Talibān-e-Pakistān in Urdu—(known as TTP in English)—formed in 2007 to unite thirteen disparate groups fighting NATO forces and to establish Islamic law in Pakistan (Harpviken 2012).[6] Since 2010, the neo-Taliban has contacted Uzbek and Tajik groups through clerical networks to rally against foreign rule, but only with limited success given the Taliban's harsh rule over minorities before 2001 (Giustozzi 2010).

The Afghan neo-Taliban has embraced modern technology, such as the Internet, email, and Twitter. Prior to 2001, the only two Internet connections in Afghanistan were in Mullah Omar's office and in the Foreign Ministry, in compliance with the Taliban's ban on human images (E. Rubin 2006). The Taliban now uses these technologies aggressively to disseminate official statements and claims of responsibility after attacks (Nordland and Sukhanyar 2013). The audience for the Taliban's electronic communications is not the vastly rural Afghan population but the group's savvy military commanders, Afghans abroad, potential donors, journalists, and opinion leaders (Chroust 2000; Nathan 2009). Consider this statement released by the Taliban on the anniversary of the 9/11 attacks in 2012 in the Urdu periodical *Sharī'at*: "The occupying Americans, who eleven years ago on the pretext of this (9/11) announced a war against Islam and Muslims, have lost their strength with the assistance of God (may He be exalted) and the tyrannical militants have subsequently retreated from other battlefields" ("Nāīn Alavan Kī Gyārahvīn Barsī Kī Bābat Imārat-e Islāmīya Kā Elāmīya" 2012). The same statement "sends a message to American rulers, their allies, and their populations to stop additional bloodshed of oppressed Afghans under the cover of 9/11 and to work with wisdom and enlightenment rather than savagery and barbarity" ("Nāīn Alavan Kī Gyārahvīn Barsī Kī Bābat Imārat-e Islāmīya Kā Elāmīya" 2012). American news sources estimated the Taliban's core strength in 2014 at sixty thousand soldiers, compared to two thousand soldiers until about 2004, with Taliban casualties ranging between twenty thousand and thirty-five thousand (Dawi 2014). Over a decade of war and the slated departure of American troops in 2016 make cultural analyses of Taliban texts timely and topical, so that we understand how the Taliban has successfully adapted and attracted new recruits. How can scholars in the social and behavioral sciences contribute to this understanding at a practical level for military and policy relevance while remaining faithful to the academic enterprise of theory building? These are the issues I explore in the next section, when considering the potential contributions of the behavioral sciences.

USING PSYCHIATRY AND PSYCHOLOGY TO STUDY THE TALIBAN

In this book, I build on theories and research methods from psychiatry, psychology, and South Asian studies to analyze official Taliban discourse through written and multimedia texts. Many psychiatrists and psychologists interested in political violence have worked within the subdiscipline of political psychology. Political psychology employs qualitative and quantitative methods to study interactions between politics and psychology, especially political movements, styles of political leadership, and group conflict (Van Ginneken 1988; Ward 2002; Deutsch and Kinnvall 2002;

Jost and Sidanius 2004). All of these areas are applicable to the Taliban. Although political psychology initially drew on anthropology to situate research findings, the cultural focus disappeared once political psychologists found certain anthropological trends irrelevant, such as the tendency to study small, distant, and ethnically homogeneous societies outside of North America and Europe, or the tendency to treat culture as a mere extension of personality (Renshon 2002).

Recently, political psychologists have corrected this imbalance by reincorporating cultural theories and models into research. To ignore culture is to ignore the social contexts of politics, such as the shared meanings of individual and collective identities, boundaries between competing groups, interpretations of political actions, and political ideologies occurring in a specific time and place (Ross 1997; Spears and Klein 2011). Culture assumes even greater importance as the globalized, networked world decreases the primacy of inherited identities such as ethnicity, race, religion, nation, and gender, and people create identities through new attachments throughout life (Nesbitt-Larking and Kinnvall 2012). Although numerous definitions of culture exist (Pye 1997), one clear definition used in this book is "the range of shared understandings and expectations that are embedded in individual psychology, societal institutions, and public practices" (Renshon and Duckitt 1997, 233). These understandings and expectations can be group discourses in society from which people create individual narratives and prescribe ways of thinking, feeling, and behaving through use of language (Foucault 1980; Foucault 1994; Renshon and Duckitt 1997; Hammack and Pilecki 2012). In a globally connected world, a cultural approach to political psychology affirms that understandings and expectations are dynamic and created, not static or immutable. Cultural analysis can determine the content and circumstances governing the creation of group discourses, such as those constructed by the Taliban in textual form and broadcast in official periodicals and the Internet as public practices.

Just as political psychologists have paid greater attention to culture, cultural psychiatrists have called for greater attention to politics. In a globalizing world, cultural psychiatrists must recognize that different political systems cultivate different self- and social identities (Kirmayer and Minas 2000). People now create individual, hybridized identities based on an unparalleled exposure to various types of transnational information (Bibeau 1997; Aggarwal 2012). The Internet acts as a hub of cultural production, revolutionizing individual and group identity through self-presentation on personal websites, social positioning in online communities, and relating to others by linking to other websites (Kirmayer 2006, Kirmayer, Raikhel, and Rahimi 2013). The Internet also reconfigures thought patterns (cognitions) and sensory experiences (affects) that shape traditional areas of research, such as the role of religion in justifying violence; the impact of war on displacement and belonging; and the effects of cultural change on individual identity (Bhui and Ibrahim 2013). In sum, cultural psychiatrists and political psychologists have concluded that an interconnected world exposes people to infinite cultural media with the potential to produce new individual and group identities bearing political implications.

These insights produce a series of interrelated research questions that orient this book: How does the Taliban strive to produce novel forms of identity and community through the Internet? What types of cognitive and affective experiences do the Taliban elicit through texts on violence, war, and cultural change? How does language frame the Taliban's multiple messages? To answer such questions, the objectives of this book are threefold: (1) to establish a general framework for analyzing multilingual media production from militant Islamist groups; (2) to apply this framework to the Taliban, with discourse analyses of texts in English, Arabic, Dari, and Urdu; and (3) to test how findings can reformulate psychiatric, psychological, and cultural theories.

THE VIRTUAL EMIRATE AS THE TALIBAN'S IMAGINED COMMUNITY

Scholars in Islamic studies have begun to study the Internet's relationship to Muslim identity. Terms such as the "digital umma [community]" (Bunt 2000), "Dar [House of] al-Cyber Islam" (Kort 2005), and the "virtual ummah" (Roy 2010) highlight how Muslims have developed transnational Internet communities outside geographically restricted social institutions, such as families or mosques. Personal home pages allow people to represent themselves anew, while links to other websites endorse group affiliations around what it means to be Muslim (Roy 2004). The Internet has also led militant groups such as Al Qaeda to subvert traditional religious authorities, such as local clerics, by propagating theological interpretations of jihad that disseminate beyond the jurisdictions of law enforcement agencies

(Bunt 2009). Some have speculated that the "digihad" propagates holy war over the web (White 2012) and that the "virtual caliphate" is based on "a complete rejection of the current world order," which is "driven by the belief that the nations of the world are not being ruled as they should be, according to a strict interpretation of Islamic law" (Lappin 2011, 4). In this sense, the Internet acts as a medium through which militant groups can reject the current world order in favor of a virtual world that they create. This scholarship has added to our understanding of the Internet's role in stimulating new individual and group identities, but has yet to consider militant Islamist movements that challenge state sovereignty.

Cultural psychiatrists (Bartocci and Littlewood 2004; Kirmayer, Raikhel, and Rahimi 2013; Kirmayer, Rousseau, and Guzder 2014) and political psychologists (Hammack 2008; Stern 1995; Ross 1997; Ross 2001; Wright 2011; Schatz and Lavine 2007; Liu, Sibley, and Huang 2013) have turned to Benedict Anderson's (1991) theory of the "imagined community" to explain group identity formation through communication. We can use this theory to contemplate the Taliban's success in marshaling support. Anderson defines the nation as "an imagined political community—and imagined as both inherently limited and sovereign" (6): *imagined*, since most members of the nation will never meet one another; *limited*, since this community has boundaries and is not all of humanity; and *sovereign*, since all nations dream of self-rule (6–7). Anderson posits that the following social processes increase nationalism. With the founding of the modern nation-state as the highest source of power within society, "state sovereignty is fully, flatly, and evenly operative over each square centimetre of a legally demarcated territory" (19). Print, specifically the novel and the newspaper, provides the "technical means" for representing the imagined community of the nation to its members (25–36). Print languages connect people by creating a medium for texts that can be mass-distributed and read in simultaneity, unlike older genres, such as manuscripts, that were limited in distribution to monarchs and elite audiences (44–46). Printing increases interest in literacy, fueling popular support for nationalist movements, "with the masses discovering a new glory in the print elevation of languages they had humbly spoken all along" (80). Anderson qualifies "state nationalism" as nationalism emanating from the state for state interests (159).

I concede Anderson's point that the political community mediated by the Taliban over its web pages is also imagined, as all website users or periodical readers are unlikely to ever meet one another. However, other arguments about sovereignty do not hold. First, despite strong political, economic, and military support from international powers, Afghanistan's government does not exert sovereignty that is fully, flatly, and evenly operative over each square centimeter of territory. The Taliban's shadow governments prove this point by exerting law and order functions that compete with state institutions. At the same time, the Taliban's shadow governments do not operate fully, flatly, and evenly, since they must compete with official institutions. Neither the government nor the Taliban can claim control over every square centimeter of Afghan territory.

This situation forces us to rethink how nationalism, territorial sovereignty, and personal identity interrelate in the Taliban's formation of a political community. It has been suggested that the unity of the nation-state breaks down in times of political violence, when the state, defined as "that ensemble of discourses and practices of power," works against the nation, defined as "the people," to subjugate internal enemies (Aretxaga 2003). The Taliban represents itself as championing the people against the official government's practices of power. The editor of the Taliban's Urdu monthly *Sharī'at* emphasizes this point:

> The fundamental goal of the monthly "*Sharī'at*" is not just providing the truth and presenting the real picture to people, but also stopping and starting to remedy the wily enemy's false propaganda. You are very well aware of the fact that in the Islamic Emirate of Afghanistan over the past ten years, there is no example in the recent past of the way that the holy warriors have forced the occupying enemy to eat dust, and there is no day that passes when the caravans of the enemy do not face strikes from attacks of the holy warriors. (Afghan 2012, 2)

Afghan (2012) presents the Taliban on the side of the nation, against the Afghan government, which is equated with the state. The editor uses positive words, such as "truth" for the Taliban's "holy warriors," who have sided with "the people" against the "false propaganda" of an occupying "wily enemy" allied with the Afghan government. This Taliban text, along with previously mentioned texts on the Taliban's bureaucracies, forces us to rethink theories on the relationship between nation and state.

The Taliban also compels us to revisit the relationship between print and linguistic nationalism and political community formation. Novels and newspapers allow authors to represent the imagined community to its members through media in simultaneous mass distribution. Anderson (1991) predicts that the future will bring new forms of nationalism: "Advances in communications technology, especially radio and television, give print allies unavailable a century ago. Multilingual broadcasting can conjure up the imagined community to illiterates and populations with different mother-tongues" (135). We can add the Internet to this collection of new communications technologies. However, Anderson assumes that print stimulates the masses to find "a new glory" in languages they had "humbly" spoken, implying that the imagined political community crystallizes through languages belonging to a dominant ethnic group. This connection between language and nation roots the idea of a state *anthropologically*—through an ethnically, culturally, and linguistically homogeneous population—rather than *politically*—as a group of people who contract to live under common laws, irrespective of culture, language, or ethnicity (Hobsbawm 1996).

Anderson's claims for linguistic nationalism may have been true for the Taliban and the Pashto language before 9/11, but they do not explain the Taliban's subsequent writings in Arabic, Dari, Urdu, and English. No ethnic group in Afghanistan speaks Arabic, Urdu, and English natively. This point also does not hold true for Dari, whose status the Taliban downgraded during its 1996–2001 rule. A theory based on linguistic nationalism does not answer why Arabic, Dari, Urdu, or English speakers would feel a sense of attachment to the Taliban through a shared group identity as a political community.

It has been implied that the Taliban updates electronic communications in multiple languages frequently for "mobilization of Muslim opinion worldwide as a source of funding, moral support and volunteers" (Giustozzi 2008, 138). I see Taliban communications as a novel iteration within the longer trajectory of multilingual publications in Afghanistan. In 1980, Gulbuddin Hekmatyar's Hizb-e Islami began publishing periodicals in Arabic, Dari, English, Pashto, and Urdu, featuring interviews with officials, translations of renowned Islamic writers on religion and politics, religious interpretations of Afghanistan's history, quotations of the Quran and Hadith to interpret political events, and biographies of martyrs (Edwards 1995). As I show in future chapters, Taliban texts continue this tradition of multilingual production in the same languages and genres. During the 1980s, as many as thirty-five thousand Pakistanis, Arabs, and Africans joined a quarter million Afghans to expel Soviet forces through jihad, some funded by American, Pakistani, and Saudi intelligence agencies but most organized through grassroots Islamic charities featured in jihadist publications (Williams 2011). Afghan history teaches that multilingual publications have been a public practice of resistance among non-state militants.

Nonetheless, this finding validates Anderson's point about print enabling simultaneous mass distribution, not linguistic nationalism. In fact, Taliban authors disprove linguistic nationalism as a social process for a shared sense of group identity and political community. Author Akram Tafshin (2013b) underscores this point in naming obstacles to publishing in a foreign language:

> In issuing *Sharī'at*, how many difficulties were there and how these problems were solved with the cooperation of writers and editors! There was a time when the Islamic Emirate decided to issue an Urdu publication and there was only one writer available who was a little bit familiar with Urdu. And that was very little. We used to feel anxious over where to gather writers familiar with Urdu and find good writers for the Urdu periodical. For other newspapers and magazines that have full liberty in all respects, finding writers is not very difficult. (7)

The masthead of the Taliban bilingual Dari-Pashto periodical shown in figure 1.2 includes a "Subscription" section with cost by currency. The first two entries may be intuitive: a subscription in Afghanistan costs 180 Afghanis, and a subscription in Pakistan costs 240 rupees. However, the two entries below indicate worldwide circulation: a subscription in the United States is 60 dollars, and a subscription in Europe is 60 euros. We have no way of knowing how widely this periodical circulates, and it is possible that listing prices by currency is a form of propaganda, since no address for a subscription office is listed. However, the masthead indicates that the Taliban imagines itself as a global political community beyond Afghanistan.

Based on these data, I propose that we think of the Taliban, its Afghanistan campaign, and its associated websites as a *virtual emirate*. The word *virtual* carries three connotations that act as a useful heuristic for analyzing how the Taliban constructs political community. First, it calls attention

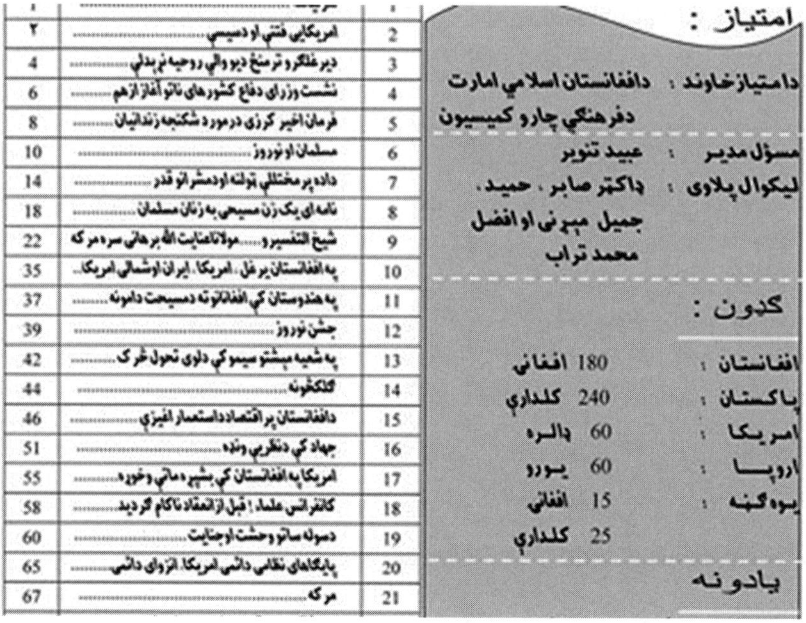

FIGURE 1.2 Masthead of the Taliban periodical *Srak*.

to the Internet as the medium for creating and disseminating shared expectations and understandings of the Taliban's community in virtual reality. Desktop "print" publishing permits the mass circulation of articles in hypertext and PDF format even sooner than the genres of books and novels that prevailed in the twentieth century. The cheap costs of Internet publishing allow Taliban authors to publish in real time and to reach a global audience beyond local borders in simultaneity.

The omnipresent, international, yet deterritorialized nature of the Internet in which the Taliban stakes its claim also requires a reassessment of how identity, space, and place relate to culture. As anthropologists Akhil Gupta and James Ferguson (1992) write: "If one begins with the premise that spaces have always been hierarchically interconnected, instead of naturally disconnected, then cultural and social change becomes not a matter of cultural contact and articulation but one of rethinking difference through connection" (8). The Internet is an interconnected space with hierarchies based on language and access, and the virtual emirate emphasizes the Taliban's articulations of cultural difference despite connections with other users. For example, in contrasting the West (*maghrib*) with the Muslim world, one Urdu author describes a YouTube video of four American soldiers who urinate on three dead Afghans:

> Will these people make us civilized? Is the name of this very thing Western civilization and progress? Are these very actions their secret to being civilized and progressive? What should be the name given to this bestiality and barbaric act? Is this the "respect" of those nations of the world that they do not tire of spreading day and night? Is playing with the honor of others the only sign of being progressive and civilized? (Balochi 2011)

YouTube becomes a site of interconnected Internet space for the Taliban to excoriate the supposedly progressive and civilized nature of Western culture. The Taliban is not decrying the existence of the Internet or YouTube, but engaging with them as spaces of cultural difference despite interconnection. A video of American soldiers urinating on dead Afghans is not assigned a cultural meaning of victory but of barbarity in an overt act of marking differences.

Second, *virtual* can also refer to a degree of incompletion, as in "nearly," "partially," or "almost," and the virtual emirate reflects the Taliban's struggle for sovereignty in Afghanistan. Unlike many militant groups around the world, the Taliban successfully managed a movement that came to domestic power and even garnered international recognition. Although NATO forces have displaced the Taliban from government positions since 2001, the Taliban has maintained shadow governments throughout Afghanistan's provinces. The virtual emirate highlights the partial, near, but ultimately incomplete nature of Taliban rule, in contrast with Anderson's conception of sovereignty as fully, flatly, and evenly operative over every portion of territory within a state's borders. Whether in the 1990s, when the Taliban occupied over 90 percent of the country, or in the 2000s, with shadow governments exercising genuine law and order functions, the Taliban have virtually, though never fully, established an autonomous emirate.

The Taliban's control over human lives in its areas has forced domestic and international governments to the negotiating table—to avert war before 2001, by demanding the surrender of Osama bin Laden, and to promote peace and reconciliation after 2001, with the departure of all American forces by 2016. If we accept one definition of sovereignty as "the exercise of control over the lives, deaths, and conditions of existence of those who fall within its purview—and the extension over them of the jurisdiction of some kind of law" (Comaroff and Comaroff 2009, 39), then we can remove the association between sovereignty and state governance. The internationally recognized Afghan government may enjoy state governance, but the Taliban enjoys sovereignty in many areas. An exploration into the virtual emirate affords us a view into the jurisdiction of the kind of law underpinning the Taliban's shadow governments. Many multiethnic societies ravaged by violence consist of "a horizontal tapestry of partial sovereignties" at war, rather than "a vertically integrated one [sovereignty] vested in the state" (Comaroff and Comaroff 2009, 39). Interviews with members of shadow governments bolster the Taliban's claims of sovereignty and act as a recruitment strategy. The first issue of *Sharī'at* includes an interview with the deputy governor (*nā'ib amīr*) of Helmand province Mullah Muhammad Dawud Muzammil, who describes activities as "very successful and effective in which the enemy suffered heavy losses and the mujahideen achieved very many victories" ("Sūba-e Helmand Ke Nā'ib Amīr Mulla Muhammad Dāwūd Muzammil Se Numāinda-e *Sharī'at* Kī Ek Nishast" 2011). As I show in chapter 4, the interviewer asks Muzammil details on the enemy's retreat from Helmand province to refute American and British accounts of the war, slamming the official government of Hamid Karzai's claims of sovereignty. The Taliban uses its partial and incomplete control to advocate for jihad and sovereignty over every centimeter of the country.

Finally, *virtual* shares an etymology with the word *virtue*, and it is on the basis of virtue and morality that the Taliban has founded the *virtual emirate*. By *virtue*, I refer to "how societies ideologically and emotionally found their cultural distinction between good and evil, and how social agents concretely work out this separation in their everyday life" (Fassin 2008, 334). Rather than imagining a political community through linguistic nationalism, the Taliban forges common identity by distinguishing between good and evil steeped in a distinct interpretation of Islam. As social agents, Taliban authors work out this separation of good and evil through texts. For example, the editorial that introduces the first issue of the Taliban's English periodical *Azan* ("call to prayer" in Arabic) justifies the need for a unified moral community based on the view that Muslims worldwide are under siege:

> In times such as ours, when truth has become difficult to see and follow, by the Grace of Allah, Azan is a humble effort at renewing the call to Tawheed ["unity" in Arabic] that has been proclaimed by the Prophets and the men of Allah throughout time. From Adam to Nuh to Ibrahim to Musa to Isa and finally to Muhammad, the call to mankind has always been the same: to be free of the servitude of created beings to the servitude of The One, The Almighty, The Law-Giver, The Creator of the heavens and the earth, Allah. It is this very call that Azan aims to invite mankind towards to. It is a call to end oppression, evil and corruption from the earth. It is not a message restricted to a particular race, land or nation; rather, it is the message of salvation for the entire mankind. All Praise is due to Allah, that the areas of Afghanistan, Iraq, Palestine, Chechnya, Pakistan, Yemen, Egypt, Libya, Syria, Mali, Somalia among others—have already witnessed the proclaimation [*sic*] of this blessed call and the caravan to glory has indeed marched forth. ("Editorial," *Azan* 1 (1), 2013)

This Taliban text advances a unique reading of theology and history in appealing to political unity among Muslims worldwide. *Tawheed* is a verbal noun from the Arabic trilateral root *w-h-d*, which implies one, oneness, or singularity. This concept—mentioned in the Quran as "Say: God is One; God the Eternal; He did not beget and is not begotten, and no one is equal to Him—was originally used to refute minor Arab deities and the Christian concept of the Holy Trinity" (Schimmel 1992). The Taliban has updated this monotheistic doctrine to envision a political community encompassing the oneness of mankind through the oneness of Abrahamic prophets ending with Muhammad. This oneness is enshrined morally to "end oppression, evil and corruption from the earth," with conflicts named around the world. The author does not "restrict" this message to any racial, ethnic, or linguistic community, but the mention of Allah clearly identifies the intended audience as Muslim.

In fact, the cultural distinction between good and evil in everyday life that the Taliban propagates as virtue can change based on language. Elsewhere, author Maulana Abdullah Muhammadi (2013) cites texts from such prominent Muslim scholars as Muhammad Qutb (1919–2014) and Ibn Taymiyyah (1263–1328) to defend the Taliban's turn away from nationalism:

> The Jews and Christians, who created this secularism (out of which stem nationalism and democracy), invented the term "nationalism" and then

> used it to divide the Muslim Ummah [community] into pieces. They destroyed the collectivism of the Muslims and planted upon them their agent "Tawagheet" (Plural for Taghut—a false god: one meaning of which is a ruler who rules by other than what Allah Has Revealed). Henceforth, they divided the Muslims on the basis of nationalism and "country-ism" by using their so-called "law-making authority." They made Muslim brothers and sisters as strangers to one another and so, Muslims began to ally and disavow, love and hate for this "nation" instead of for Allah and His Messenger. (69)

In this English text, nationalism, secularism, and democracy are purported to be devised by Jews and Christians to weaken Muslims. The nation and the country, as political units, are treated as false gods, and those who pledge allegiance to the nation and country are guilty of polytheism. The only response is to transcend nationalism through a unified political community living under a common set of Islamic laws, irrespective of culture, language, or ethnicity.

Arabic texts contain similar themes about the benefits of a worldwide community. Ahmad Mukhtar (2008), editor of *Al-Somood*, states the Taliban's intentions to publish in Arabic: "Based on media importance, the periodical *Al-Somood* began its course two years ago principally to transmit the voice of holy jihad in Afghanistan to Muslims in the world, to clarify and defend its picture. And likewise, to alert Muslims in every direction to join in this holy jihad with spirit, money, and prayer" (1). The Internet has been pivotal to *Al-Somood*'s dissemination: "The scope of its spread widened throughout the world by means of the Internet which made the possibility of reading it easier" (Mukhtar 2008, 1).

However, what explains the Taliban's forays into Dari, the language it diligently tried to suppress? In 2014, the Taliban started *Haqīqat* (Truth) for a Dari-literate audience. An editorial introduces the first issue:

> The periodical *Haqīqat* has now started production alongside other publications and periodicals of the Islamic Emirate in the domain of service to culture and educational values, to explain the truth to a nation of truth-seekers, and to the Muslims of Afghanistan in order to provide a new opportunity and greater chance for honor and freedom to all those sympathetic to the country, the scholars, those who are concerned, the wise (*pūhanwāl*), students, writers, seekers (*pūyandegān*) and to furnish a

> groundwork for greater service, especially for the Dari-speaking brothers in the country. ("Editorial," 2)

The Internet enables linguistic transformations in the Taliban's presentation of identity. Dari, the language formerly eradicated from bureaucracy, is invoked for kinship among "brothers in the country." Contrary to Anderson's (1991) claims, the imagined political community does not crystallize through Pashto, the language of Afghanistan's dominant ethnicity. Instead, the imagined political community in Dari crystallizes through religion and nationalism. Recall that Hizb-e Islami periodicals in the 1980s enabled Afghan students to understand political affairs for the first time through religion rather than see religion only as a set of practices or texts memorized from the Quran and Hadith (Edwards 1995). The Taliban continues this line of populist mobilization outside traditional institutions. Rather than tribe or ethnicity, religion and nationalism unite Dari and Pashto speakers.

If Taliban authors write in Arabic and English to reach an international Muslim audience and in Dari to reach a national Muslim audience, Urdu is used as a regional intermediary. To commemorate the second anniversary of the Urdu periodical *Sharī'at*, Urdu author Akram Tafshin (2014) clarifies that *Sharī'at* was planned for the Muslims of South Asia: "The start of the monthly *Sharī'at* happened two years ago at the insistence of the Islamic Emirate on a single publication for the Urdu speaking Muslims in the regions of Pakistan, Hindustan [India], and Bangladesh" (17). Abu Zahir, the editor of the Pashto periodical *Shahāmat*, provided another rationale for publishing in Urdu:

> It is clear as light that Urdu areas and regions have a great and sweet language. Especially in today's time, the peculiarity of the Urdu language is that countless small and large books have been written in Urdu on different topics in almost all of the Islamic sciences. For this reason, we can say that nowadays after Arabic, a large and continuing wealth (*sarmāya*) of the Islamic sciences and disciplines is protected in the Urdu language. ("Roshan Mustaqbal Kī Umīd Par" 2014)

At first, we could juxtapose these texts as contradictions. However, reading the texts closely leads to another hypothesis: What if Taliban authors draw on cultural symbols that resonate by language? For example, the Dari passage invokes "honor and freedom" to rouse nationalism among a native constituency. A Muslim population that is not from Afghanistan and is not Dari speaking may find this argument irrelevant. Conversely, Taliban authors in English deride "nationalism" and "country-ism" as a ruse of Jews and Christians to invoke unity within the worldwide Muslim *ummah*. An argument for nationalism among Arabic, English, and Urdu readers may not be as convincing as for Dari-speaking Afghan Muslims.

Instead, we can use cultural theorist Sheldon Pollock's (2006) distinction of *cosmopolitan* and *vernacular* languages to challenge Anderson's claim of linguistic nationalism in multilingual communities: cosmopolitan languages display a "transregional consensus about the presuppositions, nature, and practices of a common culture, as well as a shared set of assumptions about the elements of power," but vernacular languages "inhabited much smaller zones; the limits they confronted, or rather helped to produce, were certain cultural-political isoglosses" as languages "that did not travel—and that they knew did not travel" (19, 21). The Taliban employs Arabic and English as cosmopolitan languages that travel widely, and Dari as a vernacular language inhabiting the smaller zone of Afghanistan. Pollock (2006) draws on Greek etymology to remind us of the sense of space in these words: *cosmopolitan* from *kosmos* (world) and *vernacular* from *verna* (domestic or home born). We can add a middle position to Pollock's binary formulation. Taliban texts posit Urdu as a language for the Muslims of Bangladesh, India, and Pakistan: Urdu is neither a worldwide nor a domestic language. Here, I introduce a third term—*parochial* from *parokia* (dwelling next to, neighboring)—to describe this intermediary regional position. For the Taliban, Urdu is a *parochial* language that travels outside the home, neighboring but not worldwide, as a language parochially seen as exclusive to the Muslims of South Asia. These texts suggest that conceptions of individual and group identity can change based on audience and context, and that cultural identity cannot be reduced to language affiliation (Aggarwal 2012). Taliban authors work out the ideological and emotional separation of good and evil in future chapters by inflecting religion through geocultural tropes to construct individual and social identities differently in Arabic, Dari, English, and Urdu texts.

I shall continue to argue that the Taliban's virtual emirate is an attempt to depart from the anthropological idea of a state housing an ethnically, culturally, and linguistically homogeneous population in favor of a political idea of a state housing people who live under common Islamic virtues and laws. Cultural analysis can detect how Muslim communities use canonical, authoritative, and religious texts—such as the Quran and the Hadith—in everyday life to frame current practices, social institutions, and human conditions toward an idealized future (Asad 1993). We can use discourse analyses to probe which religious texts the Taliban invokes—and for which purposes—in propagating shared understandings and expectations that mark clear online boundaries of cultural difference. Such texts are transmitted throughout a globalized, networked world to decrease the salience of inherited markers of identity, such as language, ethnicity, race, citizenship, and country of origin. Instead, the Taliban uses the Internet to great advantage in transforming individual and community identities through deliberate ways of self-presentation, social positioning, and relating to others that realize and reflect cultural change.

CHAPTER 2

Mullah Omar's Leadership in the Virtual Emirate

DECLASSIFIED INTELLIGENCE DOCUMENTS have displayed an ardent interest in Mullah Omar's leadership from the Taliban's inception. An American cable from 1994 speculated on the Taliban's origins after it wrested control of the town of Spin Boldak: "A group of fighters calling themselves the Taliban ("Seekers," a term applied to religious students) were active in the Kandahar region under Khalis [a mujahideen group] Commander Mullah Yar Mohammad. In early summer, however, another group using the Taliban name surfaced in Kandahar province under the leadership of a Mullah Mohammad Omar and his deputy, a Mullah Ehsanullah" (U.S. Consulate [Peshawar] 1994). A cable from 1995 centered on Mullah Omar's charismatic presence in the madrasa of a trader named Haji Bashar:

> Mullah Omar went to Haji Bashar and related a vision in which the Prophet Mohammed had appeared to him and told him of the need to bring peace to Afghanistan. Haji Bashar believed Mullah Omar and drawing upon family resources and local business and political connections (including the bazaaris [merchants] and Jamiat commander Mullah Naqibullah), raised 8 million Pakistani rupees (USD 250,000) for the cause and contributed six pick-up trucks. (U.S. Embassy [Islamabad] 1995)

By 1997, American officials discovered more information. His "highly personalized" leadership "weakened" Taliban institutions based on his "obscurantist" personality: "Omar sees himself as a religious figure of historical importance and he prefers to pronounce only on what he considers the most important issues" (U.S. Embassy [Islamabad] 1997). An update affirmed that "influence in the movement can basically be judged by one's closeness to Omar" and that his legitimacy rested on "four pillars": (1) piety, (2) effective command against the Soviet occupation in the 1980s, (3) opposition to corrupt mujahideen leaders after the eruption of anarchy in 1992, and (4) success in leading the Taliban to control 85 percent of the country (U.S. Embassy [Islamabad] 1998b). A briefing from the U.S. Defense Intelligence Agency two months after the 9/11 attacks listed the few facts known about him:

- He was born in Noori village near Kandahar in 1960 to a religious Pashtun family.
- He was an only child and lost his father in childhood.
- He was wounded in the eye and leg after joining the mujahideen group Hizb-e-Islami (the Islamic Party) following the Communist coup of 1978.
- He founded the Taliban in 1994 after unscrupulous mujahideen commanders abducted a woman in Meiwand district.
- He first married at the age of thirty-one, has four wives, lost one son from NATO air strikes in October 2011, and has a daughter who may have married a son of Osama bin Laden.
- He has met few non-Muslims.
- He enjoys launching rockets and driving automobiles.

Figure 2 .1 is reportedly a picture of Mullah Omar in the public domain. A decade later, journalists know little more due to his secrecy. Few have seen him, even when he dominated Afghanistan from 1994 to 2001 (Onishi 2001). He refuses most visitors, speaks little, and has been photographed maybe three times (Weiner 2001). He travels with dozens of bodyguards in a long convoy of sport-utility vehicles shielded by tinted windows (Goldenberg 1998). Despite infrequent public appearances, he remains lionized among the Taliban. At its founding in 1994, "each man took an oath of loyalty to Omar," after which "no senior Taliban commander has ever betrayed [his] whereabouts" (Rashid 2010). He triumphed with thirty students in battle after hanging a rival commander accused of raping women from a tank barrel (Bartholet 2001; Rashid 2000). He has been "generous with favors," dispensing cars to covetous commanders (Johnson and Thomas 2002), and lives on a simple diet of tea, fruits, and nuts (Weiner 2001). Journalists have marveled at his simplicity, such as sitting on the floor, serving food to others, and forgoing protocol (Bergen 2006).

FIGURE 2.1 Mullah Omar.

Mullah Omar's devotion has also inspired Al Qaeda followers, such as Osama bin Laden himself. Bin Laden donated $3 million for the Taliban to conquer Kabul from enemies in 1996 (Bergen 2001). That year, Mullah Omar proclaimed himself the *Amīr ul-Muminīn* (Commander of the Faithful) of the Muslim community worldwide to silence internal rivals, rouse troops, and solicit recognition from Muslim countries (Strick van Linschoten and Kuehn 2012a). Bin Laden accepted his new title, gifting cars, weapons, and infrastructure to flatter him and secure refuge (Burke 2003). Relations were initially cordial: Mullah Omar appreciated bin Laden's support to the mujahideen against the Soviet Union, and bin Laden pledged allegiance to Mullah Omar (Atwan 2006). In 1997, Mullah Omar invited bin Laden to Kandahar for protection (Gunaratna 2002) and described him as one of Islam's most important leaders (Coll 2004). In 1999, Mullah Omar dispatched Taliban officials to protect bin Laden, his family, and senior Al Qaeda operatives after the United States vowed to retaliate for bombed embassies in Kenya and Tanzania (Bodansky 1999). Before 9/11, bin Laden reaffirmed Mullah Omar's exalted position in a speech to rally volunteers and finances for the Taliban: "On this occasion I assure you and Muslims across the world that I submit to God on the duty of allegiance to Mullah Muhammad Omar, and that I have taken my oath of allegiance to him (Lawrence 2005, 98).

Nonetheless, Mullah Omar exacted obedience that bin Laden did not always command. Members of Al Qaeda were banned from giving interviews or attacking the United States (Brown 2010). Journalists who met Mullah Omar before 9/11 were "stunned" at his hostility toward bin Laden, leading to "confiscating his [bin Laden's] cellphone, putting him under house arrest and forbidding him to talk to the press or issue fatwas [religious edicts]" (Atran 2010). Internal memos from Al Qaeda urged bin Laden to obey Mullah Omar: "The Leader of the Faithful was right when he asked you to refrain from interviews, announcements, and media encounters, and that you will help the Taliban as much as you can in their battle, until they achieve control over Afghanistan" (Cullison 2004). Mid-level Taliban officials tired of Al Qaeda's heavy-handedness and transnational aspirations, but feared reprisals for voicing disagreements (Strick van Linschoten and Kuehn 2012). Shortly after George Bush assumed the presidency in 2001, American diplomats met with the Taliban over concerns that bin Laden was planning attacks on the United States, an allegation that Mullah Omar denied (Zaeef 2010). After 9/11, Mullah Omar convened a council of clerics to decide whether bin Laden should be surrendered to the United States, tried in Afghanistan, or tried in a third country with an Islamic legal system, but he chose to house bin Laden indefinitely after the United States insisted on extradition (Atran 2009). Disagreements among Taliban and Al Qaeda officials did not stop the Islamic Emirate from eulogizing bin Laden upon his death:

> The Islamic Emirate extends its condolences regarding this great tragedy to the entire Muslim *ummah*, to the family of the martyr, to his followers, and to all mujahideen. And asks God (may He be exalted) to accept the

> sacrifices of the martyred Sheikh and to rescue the Islamic *ummah* from the current tragic position with the blessing of jihad and his martyrdom on His path.
>
> Indeed, the martyr—may God (may He be exalted) have mercy upon him—was a supporter of the Afghan Islamic jihad against the antagonist Soviet Union and joined with all sincerity and bravery in the jihad with the Afghans until the invading Soviet forces departed from Afghanistan. And he advanced in jihad in the path of God with great sacrifices of which the history of the Islamic *ummah* will be proud forever. ("Bayān Al-Shūra Al-Qayādī L-il Imāra Al-Islāmīya Bi-Munāsaba Istishhād Al-Mujāhid Al-Kabīr Al-Shaikh Al-Shahīd Usāma ibn-e Lādin—Rahama-hu Allah Ta'āla" 2011)

Although the 9/11 attacks have scattered the Taliban and Al Qaeda, Mullah Omar's persona has remained large. A Taliban official in Pakistan declared that "everything in the Taliban starts and ends with Mulla[1] Omar's orders" (Shahzad 2010). Legends abound about him crisscrossing the countryside on motorbike to rally troops (Yousafzai and Moreau 2004). He embraced smart phones to respond quickly to queries from the Taliban ruling council responsible for implementing his decisions (Yousafzai and Moreau 2007). He also sought a softer image by pledging "to maintain good and positive relations with all neighbors based on mutual respect" (Rid 2010) and banning Taliban fighters from carrying out suicide bombings, burning buildings, and dismembering enemies (A. Rubin 2010). Still, years of living incognito have led soldiers to question whether he is still alive or whether his legend has been manipulated (Yousafzai and Moreau 2010). In July 2015, the Taliban released an official statement that Mullah Omar had passed away, but it waited until September 2015 to clarify that his death occurred in 2013, leading to deputies running the organization (Islamic Emirate of Afghanistan 2015a, 2015b).

Hence, rumors of Mullah Omar's demise prompted Taliban authors to publicize messages allegedly authored by him to forestall demoralization. Before news of his death, it had been conjectured that the Taliban exploited his name: "The Taliban's propaganda chief liked to have Mullah Omar's name on every communiqué—'to make it more authentic and reliable'" (Seibert 2012). I believe that his authorship of these texts is secondary: for two years after his death, the Taliban profited from his image to recruit insurgents. This chapter investigates shared cultural understandings and expectations around Mullah Omar's leadership. First, I survey extant work on his leadership. Cultural psychiatrists and psychologists have written extensively on cross-cultural measurement, and this scholarship may point to new directions in the political psychology of leadership. Next, I analyze texts authored by Taliban authors in the name of Mullah Omar. I apply what I term a *psycho-cultural assessment of political leaders* to follow psychological and cultural themes. Finally, I consider why his persona may have marshaled followers in the deterritorialized world of the virtual emirate.

STUDYING THE LEADERSHIP OF MULLAH OMAR

There is no consensus on approaches to cross-cultural research in psychiatry and psychology, including studies on leaders from other cultures. Three schools of thought are relevant to political psychology leadership profiles based on personality traits:

1. Some (McCrae 2000; Hofstede and McCrae 2004) believe that research on personality traits, begun in North America and Europe, can be applied across cultures, since genetic studies prove that personality traits are inherited. In this view, culture merely shades their expressions. Therefore, personality trait measures developed with North Americans and Europeans can be used with populations elsewhere.
2. Dissenters (Shweder 1991; Shweder and Sullivan 1993) point to anthropologists and psychologists who show that culture and psychology constitute each other from childhood—we learn ways of thinking, behaving, and feeling upon becoming members of societies, no two of which are equal. Rather than assuming the universality of personality traits across cultures, researchers should first describe personality traits prevalent within each culture.
3. An intermediary position (Church 2000, 2008; Triandis and Suh 2002; Lehman, Chiu, and Schaller 2004; Heine and Buchtel 2009; Cheung, van de Vijver, and Leong 2011) advocates using measures with assumed universality to compare traits across cultures and descriptions of trait uniqueness within each culture. Measures may need revision if their dimensions, study populations, and conclusions are not cross-culturally valid. I adopt this

position to identify what makes Mullah Omar both unique and similar to other leaders.

Psychological studies of leaders have mostly used two methods: *psycho-biographical case studies* developed by psychiatrists and psychologists, and *textual content analyses* developed by political scientists and social psychologists (Post, Walker, and Winter 2003). I summarize each and consider their possible applications to Mullah Omar.

Psycho-biography and the Political Personality Profile

Jerrold Post's political personality profile exemplifies government interest in psycho-biographical studies of leadership. A psychiatrist, Post developed the psychological profiling of world leaders for senior officials, such as the president, the secretary of state, and the secretary of defense (Post 2003a). An article from the Central Intelligence Agency's in-house journal details his method:

> In addition to drawing on all classified reporting, a thorough review of the open literature is conducted. Official and unofficial biographies have often provided key background material and insights, as have television, newspaper and magazine profiles. When there are significant holes in the data, requirements are sent to the field to attempt to develop the missing information. (Post 1979, 3)

This method of distantly assessing political leaders has inspired the few studies we have on Mullah Omar, so I cover it in depth.

Post has adapted the psychiatric case study to connect life events with a leader's beliefs on leadership, power, behaviors, worldview, and personality. He defines personality as "a systematic pattern of functioning that is consistent over a range of behaviors and over time" (2003a, 69), concentrating on formative events, management and negotiating styles, decision making, and rhetorical style. Because personality represents a consistent pattern of functioning, personality types can be derived from comparing similar responses to different events (78–82). Post relies on the fourth edition of the American Psychiatric Association's *Diagnostic and Statistical Manual of Mental Disorders* (*DSM-IV*) to describe three prototypes of leaders: the narcissist, the obsessive-compulsive, and the paranoid (82–100). Post accepts *DSM-IV*'s definition of personality disorders: "deeply ingrained, inflexible, maladaptive patterns of relating to, perceiving and thinking about the environment and oneself that are of sufficient severity to cause either significant impairment in adaptive functioning or subjective distress" (American Psychiatric Association 1994, 630). Each personality disorder is associated with a leadership style, though "most individuals, and most leaders, possess a broad array of characteristics that do not fit one pure type" (Post 2003a, 100). The profile intends "to characterize the core political personality, identifying the deeply ingrained patterns that are coherent and accordingly have powerful predictive implications" (69).

Since *DSM-IV*, cultural psychiatrists have questioned such assumptions about personality and personality disorders. Definitions of personality normality and deviance vary across cultures based on values, ideals, and social structures (Alarcón and Foulks 1995a). *DSM* has been criticized for entrenching American cultural values, since the studies underpinning its diagnostic classifications have mostly come from the United States (Littlewood 1992), namely upper-middle-class Caucasians who enrolled in research at American academic centers (Good 1996). Prevalence rates of the *DSM-IV* personality disorders have been validated in few Asian countries (Tyrer et al. 2010), with no studies from Afghanistan. Frequent revisions of the *DSM*—now in its fifth edition as of May 2013—also threaten the use of one edition (such as *DSM-IV*) for basing future psychological assessments of political leadership.

Moreover, the dilemma of composing political personality profiles for leaders from cultures other than our own rests in explicating how these cultures comprehend the self, responsibility for behaviors, and relationships. For example, *egocentric* societies (such as the United States) define the self individually, idealize values of autonomy and achievement, and presume that personality traits are inherent and enduring; *sociocentric* societies define the self through family, clan, or community; idealize values of collectivism, honor, and piety; and presume that personality traits can change by situation (Kirmayer 2007). Knowing that *DSM* personality disorders enshrine American egocentric values of individualism and autonomy (Littlewood 1993, 1996) and that Afghanistan is a sociocentric society based on tribe (Rasanayagam 2003; Barfield 2005, 2010; Haroon 2007), we should proceed cautiously in applying *DSM* categories to Mullah Omar.

I do not claim that knowledge produced in the United States is unsuitable elsewhere. To dismiss psychological theories would be like dismissing economics, history, anthropology, and other sciences from North America or Europe, when the emphasis should be on testing and reformulating theories elsewhere (Obeyesekere 1990). Instead, this knowledge should be cross-culturally valid. Studying personality across cultures comes with the possibilities of misinterpreting behaviors based on cultural unfamiliarity. Culturally valid studies distinguish between cultural and psychological explanations of personality function and dysfunction (Alarcón and Foulks 1995b). Therefore, we should determine whether the individual and cultural group view behaviors under question as normal, and whether these behaviors come from social circumstances rather than the self (Fabrega 1994). Post (2003a) recognizes the role of culture in a leader's biography: "The importance of that political and cultural context cannot be overestimated" (71). Unfortunately, cultural analysis has not been applied in the few studies on Mullah Omar's leadership. I conducted a literature review in Google Scholar for the phrase "Mullah Omar" paired with "psychology" or "psychological" in August 2014. I read all titles and abstracts of the first two thousand results in English. The works that addressed leadership in the Taliban, covered by the following psychological topics, confirm the need for cultural analysis:

Dream interpretation. Dream interpretation has been part of Islamic tradition since medieval times (Lamoreaux 2002; Green 2003; Edgar and Henig 2010; Edgar 2011) and in psychology since Sigmund Freud's *The Interpretation of Dreams* (1913). Some contend that Mullah Omar's leadership stems from his dreams. One anthropologist has written that Mullah Omar "seems to have gained his charismatic leadership status largely through his dream accounts" (Edgar 2006, 269). This conclusion comes from a statement to a BBC reporter who interviewed Mullah Omar:

> The whole project was maybe built on this dream, he had this task or duty to perform and he must lead his Taliban, his fighters, and he must restore order and peace and enforce *Sharia*, Islamic law. . . . I was told by so many Taliban leaders, commanders, fighters; look, you know, Mullah Omar is a holy man and he gets instructions in his dream and he follows them up. The genesis of the Taliban Islamic movement was this vision, this night dream that Mullah Omar had. (Edgar 2006, 268)

Drawing on this interview, another observer posits that dreams yield insights into leaders:

> I hope the examples referred to here may be enough to alert us all to the deep unconscious processes affecting national and international conflict in our time. Psychotherapists and counsellors need to be aware that our dreams reflect not only intra-psychic personal concerns, but also the tides and currents in our society, which may be calling for social or political action. (Bowater 2012, 53)

Mullah Omar's dreams are seen as "charisma" or "intra-psychic personal concerns," with no analysis of their cultural meanings in Islamic or Afghan contexts.[2] We know little about Mullah Omar's dreams, and without knowing how they have been interpreted among his followers, we cannot speculate that he invokes them for leadership.

Followership. Inverting the interest in leaders, some have speculated on the psychological drives among the Taliban's followers. Kinnvall and Nesbitt-Larking assert:

> Those for whom the ego remains underdeveloped are pulled between two competing poles, those of the rigid superego and the unrestrained id. Muslim orthodoxies offer a range of readily available "laws of the Father" grounded in selective readings of the Qu'ran and archaic and partial applications of Shariah law.[3] An overdependence on a rigidly imposed, improperly understood, and external set of edicts leads to fundamentalism. When it is combined with periodic eruptions of volatile forces from a repressed id, it can result in forms of authoritarian behavior that may include both passive compliance and violent episodes. (2011, 65)

In this passage, followers of the Taliban are portrayed as pathological individuals "for whom the ego remains underdeveloped" and who oscillate between "the rigid superego and the unrestrained id," implying an irrationality. The authors dismiss "selective readings of the Qu'ran and archaic and partial applications of Shariah law" that lead to "fundamentalism." No details are offered of such "selective readings" or "archaic and partial applications" from the Taliban's cultural perspectives, even though senior Taliban leaders have graduated from madrasas and may be more qualified to perform religious exegeses than the authors would admit. As shown in chapter 1, the certification of the Taliban as a legitimate Islamic movement from the Grand Mufti of Pakistan challenges these authors.

A Political Personality Profile of Mullah Omar. Lanham's (2011) profile of Mullah Omar is the sole example applying Post's (2003a) method fully. Lanham acknowledges Post's influence: "Jerrold Post's outline for assessing leaders from a distance provides the basis for conducting this leadership profile and assessment of Mullah Omar." Lanham (2011) traces Mullah Omar's leadership to his self-proclamation as Commander of the Faithful:

> Moments like these capture the essence [of] the "mirror-hungry" charismatic leader "ideal-hungry" follower relationship.[4] In this case, however, unlike "reparative charismatic" leaders,[5] this "destructive charismatic" leader uses "absolutist polarizing rhetoric, drawing his followers together against the outside enemy." According to Post, "either-or, all-or-nothing, categorization is the hallmark of evocative rhetoric." In addition to this Manichean tactic, invoking divine guidance gives the leader supernatural charisma.

Lanham muses on Mullah Omar's family dynamics: "It does not take a psychologist to recognize that Mullah Omar's youth without his father was no doubt a trying period for him. Likewise the loss of any siblings and children to 'natural' causes no doubt also had their effect on Mullah Omar's psyche." Lanham traces Mullah Omar's paranoia to the NATO invasion of 2001: "Clearly he has become more concerned with his own safety over time, to put it mildly, or even borderline paranoid." Paranoia has altered his leadership: "Indeed, even in his paranoia of late, and perhaps more because of it, he is increasingly delegating the wielding of power to his second in command, [Mullah] Berader and the Quetta Shura" (Lanham 2011).

These passages analyze Mullah Omar's biography from psychological theories without testing their cultural validity, risking inaccurate interpretations. For example, Mullah Omar's assumption of the title of Commander of the Faithful is viewed as destructive rather than reparative. American audiences may see his title as grandiose, but that does not answer why it has energized thousands of Afghans. Recalling the definition of culture in chapter 1, we do not yet know the range of shared understandings and expectations embedded in social institutions such as this title. Analyzing Mullah Omar's family, Lanham (2011) assumes that a youth without a father was "trying," even though anthropologists have shown that family structure and role expectations are culturally determined (Cohler 1992; Crapanzano 1992). We know little about the nature of actual relationships in the absence of empirical evidence. Finally, using the word *paranoia* to label Mullah Omar's leadership post-2001 is unapt, since he has gone underground to evade capture. Evading capture would not be paranoia—it would be rational self-preservation. Cultural analysis can fill these gaps in psychological theory.

Textual Content Analyses of Interviews and Speeches

For some leaders, biographical data may be unavailable. In these instances, political psychologists employ interviews and speeches to detect personality traits through content analysis. Margaret Hermann (1980) writes that leaders of "closed or dictatorial political systems" can "leave traces of themselves in their speeches, interviews, and writings" (332). To classify leadership styles of Soviet politburo members, she scanned transcribed speeches and interviews in 250-word fragments (12,500 words per member) for attitudes suggestive of personality traits. She analyzed each fragment by topic, audience, and spontaneity, defined as "abrupt changes in thought, reactions to the applause of the audience, and references to personal experiences," since interviews are less contrived than are speeches (Hermann 1980, 335). She rated each leader on eight traits: (1) *nationalism/ethnocentrism*, (2) *belief in one's ability to control events*, (3) *need for power*, (4) *need for affiliation*, (5) *conceptual complexity*, (6) *self-confidence*, (7) *distrust of others*, and (8) *task orientation* (Hermann 1980, 399). Hermann (2003) has since stated that analyses can be conducted from fifty interview responses of one hundred words or more (180). She has combined *nationalism/ethnocentrism* and *need for affiliation* into *in-group bias*, which is the percentage of times in a text that a leader favorably mentions one's own group (201). All domains are rated by calculating parts of speech—such as nouns ("I") and direct objects ("me") for *self-confidence* or verbs for *belief in one's ability to control events*—for an index score (188–203). This method assumes that frequent words and phrases reflect personality traits (186).

This method also assumes that grammar is used equally across languages without cross-cultural variation, though cultural psychologists have questioned whether personality traits can be measured by counting parts of speech. We may commit *method bias* if our data, instruments, and interpretations are not cross-culturally valid (Church 2001). Hermann's (2003) method assumes that leaders use parts of speech equivalently across languages and cultures. However, North Americans and Europeans view personality traits egocentrically and intrinsic to the individual, compared to Asians, Africans, and South Americans, who view them sociocentrically and reflective of circumstances (Markus and Kitayama 1998), similar to Kirmayer's (2007) insights about personality categories across cultures. This division manifests in language use: Europeans use more adjectives to describe personality traits intrinsic to the self and enduring ("Omar is helpful") than do Asians, who use more verbs and direct objects to highlight social context ("Omar helps Osama sometimes") (Maass et al. 2006). In counting nouns and verbs to rate personality traits, we may commit method bias in assuming that Mullah Omar, a leader from a sociocentric society, uses parts of speech in the same way as do leaders from egocentric societies. This danger is especially acute since speeches, interviews, and statements at press conferences are tightly controlled, whereas personality may be determined by social context, and Mullah Omar has been underground since 2001.

A METHOD FOR PSYCHOCULTURAL LEADERSHIP ASSESSMENT APPLIED TO MULLAH OMAR

How can we construct a valid method to assess psychological and cultural dynamics of Mullah Omar's leadership? We can use both methods synergistically toward a more culturally valid assessment. The strength of Post's (2003a) method is in its rigorous presentation of diverse data and insight that behaviors can be consistently interpreted across time, gleaned from life events, and expressed in leadership. However, we need not interpret behaviors through *DSM* personality disorder categories because (1) it is not evident that Mullah Omar has a personality disorder, and (2) if he did have a personality disorder, *DSM* categories might not capture the symptoms in a cross-culturally valid manner. Hermann's (2003) framework of seven leadership traits can help, since it has been validated cross-culturally with world leaders. I am not questioning the dimensions of these traits across cultures but their measurement through grammatical analysis in light of recent psychological scholarship on language and personality measurement.

Rather than perform numerical counts of nouns or verbs, we can interpret personality traits qualitatively, since (1) measurements of personality traits through grammatical analysis have not yet incorporated linguistic research from cultural psychology, and (2) a qualitative approach allows participants to describe their meanings of the world in their own words (Creswell 2014). Qualitative research represents perspectives of participants authentically through thicker descriptions than what numbers sometimes provide (Creswell and Miller 2000; Whittemore, Chase, and Mandle 2001) and accords with discourse analysis as the methodology of this book.

Ultimately, the Taliban's secrecy imposes limitations on any assessment of Mullah Omar. Apart from a handful of interviews, all analyzed here (Laghari and Khan 1998; "Interview with Mullah Omar—Transcript" 2001; Voice of America 2001; *Al-Sharq Al-Awsat* 2002; Shehzad 2004), our only data come from recollections of close associates, his statements to the Muslim community during religious festivals, and his "orders" (*farmān*) at the end of each issue of the Urdu monthly *Sharī'at*. These orders appear in every Urdu issue, but only appear in Arabic, Dari, or English during times of religious holidays, such as Eid ul-Azha, suggesting that demonstrations of his leadership are important for Urdu speakers and reflect the Taliban's different messages by language. As of January 2015, thirty-four orders had been published; I translate the first order in entirety for readers to get a sense of the language, themes, and tone of this distinct Taliban genre:

> The attempt of our enemies is to make our young generation prisoners of an un-Islamic culture and civilization of other nations. And for this goal, those people have masterfully conducted propaganda activities. Those people are undertaking enormous amounts of financial expenditures so that our young generation can distort its Afghan and Islamic identity by any means.
>
> Oh Afghan youth! By virtue of being Muslim and Afghan, your responsibility is to gird yourself by using your every ability and preparation against the conspiracies of the enemy and to deal with the failed ruses of your historical, cultural, and religious enemy. (Islamic Emirate of Afghanistan 2012a)

From this fragment alone, we can discern that Mullah Omar identifies potential Taliban insiders as Muslims and Afghans, a point that I explore further in chapter 3. My point here is that these thirty-four orders merit close analysis for their distinct demarcations of insiders and outsiders. Apart from texts in which his name has been added to official statements, there is no other demonstration of his leadership in social media. This makes speeches and statements that are attributed to him all the more unique as a data source. Mullah Omar's English statements (Super User 2011, 2012;

Islamic Emirate of Afghanistan 2013f) alone surpass sixty-five hundred words, exceeding Margaret Hermann's (2003) standard of five thousand. I adopt the psycho-biographical case study to present Mullah Omar's behaviors as raw data (Post 1991, 2003a; Dekleva and Post 1997), complemented with content analyses of his statements by using Hermann's (2003) definition of each trait. I situate these analyses against scholarship from South Asian and Islamic studies for a clearer picture of the individual within society.

Part I: Biographical Milestones in Mullah Omar's Leadership

This section focuses on shaping events that stand out as early leadership successes and failures. The goal here is to specify sources of political identity and the influences of early experiences. A major theme running through Mullah Omar's leadership milestones is his ability to maintain composure in the midst of conflict within and outside the Taliban.

Losing an eye: An early example of leadership emerges against Soviet troops in the 1980s, reported by Abdul Salam Zaeef, the Taliban's pre-9/11 Ambassador to Pakistan: "The battle turned into a hand-to-hand fight, with grenades flying over our heads.... *Mullah* Mohammad Omar was only twenty metres away from me taking cover behind a wall. He looked around the corner and a shard of metal shrapnel hit him in the face and took out his eye" (2010, 42–43).

Leading the Taliban: Mullah Omar hesitated before accepting leadership over the militias that united in 1994 to form the Taliban: "He argued that it would be a dangerous mission, and asked us what guarantees he could have that everyone wouldn't just abandon him if things became tough. We assured him that all those involved were true *Taliban* and *mujahideen*" (original emphases) (Zaeef 2010, 64). He relented: "*Mullah* Mohammad Omar took an oath from everyone present. Each man swore on the *Qur'an* to stand by him, and to fight against corruption and the criminals" (65).

Becoming Commander of the Faithful: One Taliban official depicted Mullah Omar's assumption of the title in 1996 before Muslim clerics from Afghanistan and Pakistan: "We hadn't been told much except to come to Kandahar. On the third day, the discussion shifted and it was proposed that Mullah Mohammad Omar accepted the title of Amir ul-Mu'mineen"

(Strick van Linschoten and Kuehn 2012a, 131). Taliban mullahs believed that this title "would not imply laying claim to authority over the Muslim world as a whole, but rather to finalise and firm up the authority of Mullah Mohammad Omar and his movement, and to settle a simmering internal debate" (Strick van Linschoten and Kuehn 2012a, 131).

Destroying the Bamiyan Buddhas: Mullah Omar seethed at the international community's desire to prioritize protection of the two-millennium-old statues over Afghanistan's humanitarian needs. This neglect, rather than Islamic precepts, drove him to destroy the statues:

> I did not want to destroy the Bamiyan Buddha. In fact, some foreigners came to me and said they would like to conduct the repair work of the Bamiyan Buddha that had been slightly damaged due to rains. This shocked me. I thought, these callous people have no regard for thousands of living human beings—the Afghans who are dying of hunger, but they are so concerned about non-living objects like the Buddha. This was extremely deplorable. That is why I ordered its destruction. Had they come for humanitarian work, I would have never ordered the Buddha's destruction. (Shehzad 2004)

Protecting Osama bin Laden after 9/11: Mullah Omar justified his refusal to extradite Osama bin Laden to America since bin Laden aided the Afghan jihad:

> Osama is the greatest mujahid of the present times. He is not a terrorist as propagated by the US. He fought for Afghanistan. He saved Afghans. How could we regret hosting him? I asked the world to provide evidence again[st] him. He was innocent. Therefore, nobody—even Saudi Arabia—could prove anything against him. . . . We paid a very heavy price for this decision. But we proved that the Taliban were independent people. (Shehzad 2004)

Key relationships: Mullah Omar had several internal allies. Mullah Mohammad Rabbani commanded mujahideen from Kandahar and was Taliban deputy leader until his death in 2001 (Zaeef 2010, xxii); the Taliban have eulogized Rabbani in a "martyrdom" text that I analyze in chapter 4. Mullah Abdul Salam Zaeef fought Soviet forces with Mullah Omar and became ambassador to Pakistan until 2001 (Zaeef 2010). Haji Bashar, a former warrior of the Ittihad-e Islami (Islamic Union) party, buoyed Mullah Omar financially and became one of the world's biggest drug lords of opium (Zaeef 2010, 265). Mullah Abdul Ghani Baradar, his brother-in-law and confidant, reputedly drove Mullah Omar's getaway motorcycle after the 2001 invasion (Coll 2011; Seibert 2012).

Part II: A Personality Profile of Mullah Omar

This section identifies patterned relationships among belief and value systems, attitudes, leadership style, and other components of personality. Post (2003a) views the personality profile as crucial to understanding a leader, since personality "will often constrain the range of beliefs (or the types of belief system) that [the] individual will ultimately develop" (80). The following analyses indicate that Mullah Omar viewed himself as a divinely inspired political leader true to the tenets of Sunni Islam. He exercised complete control over matters within the Taliban while demonstrating the flexibility to adapt to external contingencies. His leadership model of a disciplined core with a strong sense of identity adapting to events as needed characterizes the Taliban's own organizational culture after the 9/11 attacks.

General description: United Nations diplomat Francesc Vendrell was one of the few non-Muslims to meet Mullah Omar, describing him as "very tall, about six foot three . . . with an enormous beard almost reaching up to his eyes. He had an empty socket where the right eye was, and wore no patch. He wore a Kandahari robe that reached just below his knees. He had bare legs and feet" (Steele 2011, 212).

Working style: Before 9/11, Mullah Omar governed Afghanistan closely, with intimate familiarity of minute details:

> Every morning from his modest office in the center of Kandahar, Mullah Omar used a satellite telephone to give instructions on how Afghanistan should be run. "He is busy with the affairs of Kabul in the mornings," said Mullah Hashim, the liaison office head. "In the afternoons he receives governors from different provinces. He is in charge of all military matters." Although the Taliban had a six-person *shura* (council), Mullah Omar, who chaired it, was more than just the first among equals. He appointed all governors, ministers and other top officials according to Hashim. (Steele 2011, 194)

Lifestyle: Before 9/11, Mullah Omar lived opulently: "There are four bedrooms for each of Mullah Omar's four wives. In one, a crib remains but there is little else. . . . There [in the barn], Mullah Omar's cows enjoyed an electric ceiling fan, lights and even individual water spigots, modern conveniences that most Afghan people live without" (Nunan 2002). He had a distinct aesthetic: "Gold-plated chandeliers hung above his bed. . . . The walls of the bedroom were decorated with moulded formica painted brown to look like wood. The private mosque was painted a lurid mixture of green and blue. The minarets even had little bits of mirror stuck to them to catch the light" (Huggler 2001). His compound sat atop an estate: "There was space aplenty—about 10 acres—within the trapezoid-shaped compound. It had clean white walls, a decorative staircase, gardens for vegetables and herbs, bathrooms with imported faucets, and out front, an elaborate bronze fountain depicting a forest with a brook" (Gargan 2001).

Health: Mullah Omar's health status was unknown before his death. The Taliban has disputed that he needed a cardiac catheterization and stent placement after a heart attack that may have caused brain damage (Farmer 2011; Stein 2011). Close associates have maintained that he suffered depression since a bomb detonated outside his compound in August 1999, killing two of his brothers (Seibert 2012). Others have confirmed that a half brother takes psychiatric medication and that his uncle used to tear off his clothes and wander the streets of Quetta naked (Seibert 2012), suggesting a family history of mental illness.

Intellectual capacity: Senior Taliban leaders in the Quetta Shura have disclosed that Mullah Omar was "a simple country preacher, without the education even to read or recite the words attributed to him, never mind actually compose them," and that "Omar used to stumble over his native tongue in the interviews he occasionally gave the BBC Pashto service when the Taliban were in power" (Seibert 2012). Unlike most Afghan mujahideen, Mullah Omar reportedly spoke "passable Arabic" (Wright 2006, 256), though an interview with an Arabic newspaper reported that he asked for written questions to be passed to his translator (*Al-Sharq Al-Awsat* 2002). His capacity for cognitive flexibility was evident in his response during a BBC interview: "The Taleban may have made some mistakes. Screening the Taleban [for loyalty] is a big task" (BBC News 2001).

Emotional reactions: Zaeef (2010) contrasts Mullah Omar's equanimity with former Afghan president Hamid Karzai's tempestuousness: "*Mullah Saheb* gave everybody who visited him enough time to empty their hearts. He listened, he was patient, and he did not react in anger. Any visitor could tell that he was thinking very deeply about what he was saying" (222). Former

Saudi intelligence chief Prince Turki Ibn al-Faisal adopted an opposite stance after a 1998 meeting on extraditing Osama bin Laden to Saudi Arabia: "He was extremely nervous, perspired, and even screamed at me. He denied that he had promised us that he would extradite bin Laden, and wanted nothing to do with a joint committee.... I could not help but think that [he] might have been taking drugs" (*Der Spiegel* 2004).

Drives and character structure: Reports are mixed on Mullah Omar's self-image. United Nations diplomat Lakhdar Brahimi met him for three hours: "Mullah Omar takes himself very seriously. He thinks he is the Amir-ul-Momineen—Commander of the Faithful. He prides himself on being frank and sincere" (Steele 2011, 209). An interview with the Voice of America (2001) after 9/11 corroborates his self-image as divinely inspired:

> I am considering two promises. One is the promise of God, the other is that of Bush. The promise of God is that my land is vast. If you start a journey on God's path, you can reside anywhere on this earth and will be protected.... The promise of Bush is that there is no place on earth where you can hide that I cannot find you. We will see which one of these two promises is fulfilled.

At times, Mullah Omar claimed divine powers: "The plan is going ahead and, God willing, it is being implemented. But it is a huge task, which is beyond the will and comprehension of human beings" ("Interview with Mullah Omar—Transcript" 2001). However, former Taliban military commanders have countered: "His mates couldn't believe it when he led the uprising in 1994.... He had always been so lacking in confidence" (Seibert 2012).

Mullah Omar derived his morality from faith. A former Taliban ambassador to Saudi Arabia noted that Mullah Omar "insisted on solving every problem in light of Sharia (Islamic law)" (Coll 2011, 49). Mullah Omar issued edicts that "no leader of a non-Muslim country should be sent a congratulatory note on his birthday" or "receive a message wishing him a long or healthy life" (Bergen 2006, 250). His reality testing was suggested by a 2001 Taliban anecdote:

> In Kandahar, Mullah Omar had resorted to fortune-telling and dream-interpretation to aid him in his decisions. He told his commanders to continue fighting, because America would soon be destroyed. But the de-

> struction of America kept all waiting while the destruction of Kandahar became a reality. Mullah Omar finally gave up dreaming and decided to abandon Kandahar. (Bergen 2006, 326)

Another recollection from 1994 situates claims about paranoia: "A report reached us that [rival commander] Sarkateb was planning to assassinate *Mullah* Mohammad Omar. He wanted to attack our leader's convoy on the road from the city back to his house. *Mullah* Mohammad Omar stopped using the road; it was no longer safe for him to travel" (Zaeef 2010, 71).

Interpersonal relations: Mullah Omar was supreme leader of the Taliban. Influential figures—former Kabul Council of Ministers head Mullah Mohammad Rabbani and former foreign minister Ahmed Wakil Muttawakil—unsuccessfully opposed his hospitality to Osama bin Laden (Strick van Linschoten and Kuehn 2012a). In 2003, members calling themselves the Jaish ul-Muslimeen (the Army of Muslims) seceded over frustrations with his leadership, but they rejoined in 2005 (Giustozzi 2008). Independent groups such as the Haqqani and Mansur networks pledged allegiance to him, a condition of remaining within the Taliban's loose confederacy (Ruttig 2009). Below Mullah Omar and the Leadership Council, four regional councils have governed local operations (Ruttig 2009).

Internationally, Mullah Omar had no allies. The Taliban enjoyed relations with the United Arab Emirates (CNN.com 2001) and Saudi Arabia ("Saudi Arabia" 2001) until refusing to surrender bin Laden following 9/11. Pakistan has operated less uniformly, with some intelligence officers warning Taliban officials of American plans before the 2001 invasion (Rashid 2001b) and others arresting members of the Taliban's Leadership Council (Filkins 2010; Gopal 2010).

Part III: Mullah Omar's Worldview

A consistent theme throughout Mullah Omar's worldview was his division of the world into Muslim insiders and non-Muslim outsiders. We saw in chapter 1 that the Taliban's originally stated goal in the late 1990s was the implementation of Islamic law throughout Afghanistan. While this goal remains a central preoccupation, Mullah Omar issued statements about events throughout the Muslim world. In the chapters ahead, Taliban authors extend this worldview in classifying in-groups and out-groups.

Perceptions of political reality: In a 1998 interview on the Taliban's website, Mullah Omar described his political outlook. Jihad organizes his political thought:

> The Islamic World either has no knowledge of our methodology, are not happy with our conduct, or they still do not have trust in us, but then the greatest point I feel is that we are still busy with the *Jihaad* in Afghanistan and as yet, have not established communication with the outside World. When with Allah Ta'ala's [God (may He be exalted)] blessings and grace we do stop the mischievous forces and establish an Islamic Rule in Afghanistan, we will establish contact with them and present our theory and standing before them and when we put before them our practical system of government, then *Insha Allah* [God willing] the Islamic World and the World's other countries will themselves give us full support. When they observe our system of simplicity, peace and security, the Islamic World will not only support us, but will also introduce our methodology in their own countries. (Laghari and Khan 1998)

We shall see in chapter 5 that this focus on "the Islamic World" contributes to the Taliban's perceptions of allies and enemies as writers conform to Mullah Omar's tendency to characterize countries by religion.

Concepts of leadership and power: Mullah Omar used his role as Commander of the Faithful to attract others. Former Pakistani Army chief Pervez Musharraf explains his frustration over the Bamiyan Buddha statues:

> When we went alone to negotiate with him on behalf of the world, we found him to be on another wavelength. He said that God wanted him to blow up the statues of Buddha because over the years God had caused rain to create huge holes at their bases where dynamite could be planted. This was a sign from the Almighty that the statues were to be destroyed. Mullah Omar paid no heed to us, and—tragically—destroyed the statues. (Musharraf 2006, 215)

Zaeef (2010) details another interaction with Pervez Musharraf by describing a letter sent by Mullah Omar in early 2001: "*Amir ul-Mu'mineen* called on President Musharraf to implement Islamic law and to give Pakistan an Islamic government. He explained the obligation of Islam and the role of an Islamic government" (120).

Domestic views: Mullah Omar's domestic priorities were clear in this 2012 statement to the Muslim community, emphasizing order, economic development, Islamic law, and Afghanistan's integrity:

> After the independence of the country, we should have a sharia-based national regime with the help of Allah, the Almighty, which will bring about a sole central authority and be free from every kind of discrimination and biases. Jobs will be given on merits; territorial integrity of the country will be protected. Security will be maintained, Sharia laws will be implemented, we will guarantee rights of both male and female of the country, build economic structures and strengthen social foundations and facilities of education for all people of the country in the light of the Islamic rules and national interests, conduct all academic and cultural affairs efficiently and will, with the help of the brave nation, foil the effort of those who harbor the notion of disintegrating Afghanistan and intend to flare up domestic war to achieve their goals. (Super User 2012)

Once again, Mullah Omar fused religious and national identities for his audience. He also promoted modern education: "The people need the new generation to be adorned with religious and modern education since the acquisition of modern knowledge is among the most important needs of restoring society in the present time" (Islamic Emirate of Afghanistan 2013e).

International views:[6] Mullah Omar routinely criticized the United States:

> From the very first day of your invasion of Afghanistan, you did not lend an ear to our stance or you looked at it with an enemy's eye but now you have seen that you have journeyed in vain for the past ten years, spent your treasure and spilled your blood in a wasted effort. How do you weigh the condition of Afghanistan and America compared with the past decade? If you do not try to delude yourselves and the world than [*sic*] indeed you will accept that both the countries have regressed, not progressed compared to that time. (Super User 2011)

Mullah Omar released a statement against Israel after its war with Lebanon in 2006, denouncing Israel and the United States: "The Israeli aggression against Lebanon expresses aggression against all Muslims, and the

United States wants to rule over the Islamic world and take its goods. So the Israeli aggression in reality is American aggression, and we appeal to all Muslims to stand in a single line against their enemy" (Islamic Emirate of Afghanistan 2006a, 2).

In 2013, Mullah Omar addressed Muslims of the Middle East, deviating from the Taliban's original objectives before 9/11 to implement Islamic law only throughout Afghanistan:

> I pray to Allah (SwT) [Blessings upon the Highest] to bring to an end, the sufferings and miserability [*sic*] of the Muslims both in Afghanistan and in the entire world, particularly, may Allah save the oppressed and believing people of Syria and Egypt who spent the (whole) month of Ramadan under beating, bloodbath, arrest and torture in squares, prisons and hospitals. (Islamic Emirate of Afghanistan 2013f)

Statements such as these raise questions as to whether the Taliban has expanded its conceptions of insiders and outsiders, themes explored in the chapters ahead.

Nationalism and country identification: Mullah Omar equated Muslim and Afghan identities, as shown in the following statement: "We want to have good relations with all those who respect Afghanistan as an independent Islamic country and their relations and approaches are not domineering and colonialist. This is what every independent and Muslim Afghan wants" (Super User 2012).

Part IV: Mullah Omar's Leadership Style

Mullah Omar demanded full obedience and did not tolerate disagreements. He expected full conformity to his vision within the Taliban, at times risking his legitimacy among followers when external events did not match his expectations.

Role expectations: Grand Mufti of Pakistan Rasheed Ahmad Ludhyanvi, who ruled that the existence of the Taliban followed Islamic law, has offered a cultural explanation of Mullah Omar's role as Commander of the Faithful:

> He holds absolute power of implementation [of Shari'a] too, so anyone who opposes him will be called a rebel according to Sharee'ah. It would then be

a fardh [requirement] to execute him. Whomsoever is called for Jihaad by the Ameer it would also be a fardh upon him to obey. (Moosa 1998)

Strategy and tactics toward followers: Mullah Omar's strategy toward followers came from the most recent code of conduct for the mujahideen:

> It is obligatory for every Mujahid to obey the leader of his group, for the leader of the group to obey the district leader, for a district leader to obey the provincial leader, for a provincial leader to obey the regional administrator, and for the regional administrator to obey the Imam or his deputy on the condition if [such obedience] is allowed under Sharia. (Shah 2012, 464)

Individual mujahideen are forbidden to speak for the Taliban: "The spokespersons of the Islamic Emirate are appointed by the leadership, after recommendation by the relevant department, to represent the entire Islamic Emirate. No one has the right to talk to the media as a messenger for provinces or groups or persons. Following this principle fully will prevent disorder, confusion, and disunity" (Shah 2012, 463).

Mullah Omar also tried to raise follower morale:

> That you are constantly ready, desirous, and prepared to present sacrifices for your religion and for the defense of the people and country—this is proof of your strong faith, lofty courage, firm determination, religious honor, taintlessness, and nobility of disposition. Oh yes! You are the cause of honor and pride for the entire world, and especially for the Islamic world. You are without a doubt from among the first ranks of freedom and self-determination and dignified champions of greatness and manliness in the twenty-first century. (Islamic Emirate of Afghanistan 2012c)

Oratorical and rhetorical style: Mullah Omar's verbal style appears lofty in his depictions of the United States as an enemy of Muslims worldwide:

> We do not consider the battle to end in Afghanistan or even in Palestine or other Muslim countries. But especially with Afghanistan, I say that the battle has begun and its forces have quickened and its forces will reach the White House because it is the base of oppression and tyranny, where they carried out an attack against Islam and Muslims without legal or international justification. And I say that this is certain with the help of God (may He be exalted). (*Al-Sharq Al-Awsat* 2002)

Elsewhere, he declared: "America is the greatest evil on earth. It is the enemy of Islam. Whoever is the US friend is the enemy of Islam. Killing the enemies of Islam is jihad. We have already consigned to hell more than 1,000 infidels that include the Americans, their allies and their Afghan flunkies" (Shehzad 2004). In both selections, the United States is characterized in stark, polarized terms.

Decision-making and implementation style: Mullah Omar made decisions unilaterally. Zaeef (2010) depicts his appointment as ambassador to Pakistan in 2000: "We had just left Kabul when I heard the announcement on the radio. As with my previous appointments there had been no discussions with Amir ul-Mu'mineen, and the nomination came as a surprise" (101). Mullah Omar once disagreed with hundreds of prominent leaders on whether bin Laden should leave Afghanistan after 9/11: "The council's meeting was intended simply to rubber-stamp Omar's opinion, but once they had announced their decision contrary to that opinion, Mullah Mohammad Omar could not renege on his promise" (Strick van Linschoten and Kuehn 2012a, 224–225).

Formal and informal negotiating style: Zaeef (2010) narrates Mullah Omar's circumspection in negotiating with Northern Alliance rival Ahmed Shah Massoud: "While he would grant Massoud a position in the political or civilian sector, he thought it would be dangerous to share power in the military. From *Mullah Saheb*'s perspective he thought that giving Massoud power over the military would create more problems than it would solve" (Zaeef 2010, 88). Zaeef (2010) also indicates how Mullah Omar's negotiating style exhibited lack of international awareness after 9/11:

> *Mullah* Mohammad Omar was unwilling to believe the details of what I had told him; he reasoned that America couldn't launch an offensive without a valid reason, and that since he had demanded that Washington conduct an official investigation, and deliver incontrovertible proof incriminating bin Laden and others in the 11 September attacks, the government of Afghanistan would take no further steps regarding the matter till they were presented with such hard evidence. In *Mullah* Mohammad Omar's mind there was less than a 10 percent chance that America would resort to anything beyond threats, and so an attack was unlikely. (Zaeef 2010, 149)

Part V: Summary and Outlook

We can now analyze Mullah Omar through Margaret Hermann's (2003) seven leadership traits. These traits support the data analyzed in this chapter, with Mullah Omar exacting submission from followers by portraying himself as religiously guided. Although he was fervently committed to dividing the world between Muslims and non-Muslims and dominating others through military means, he was also a pragmatist after the American invasion of Afghanistan in 2001.

(1) *Belief in ability to control events* is "a view of the world in which leaders perceive some degree of control over the situations in which they find themselves" (M. Hermann 2003, 188–189), with leaders high in this trait controlling decision making, compared to leaders who delegate authority (189). Mullah Omar exhibited strong belief in his ability to control events, evidenced by the appointment of officials even without their knowledge, full control over military matters, and defiance of international protest through destruction of the Bamiyan Buddhas and protection of bin Laden.
(2) *Need for power and influence* is "the desire to control, influence, or have an impact on other persons or groups," with leaders high in this trait qualified as "daring" and with "little real regard for those around them," compared to leaders who empower others and share responsibility (M. Hermann 2003, 190–191). Mullah Omar had a high need for power and influence in assuming a grand title of religious import, using his position to outmaneuver internal dissent regarding Osama bin Laden, and enforcing hierarchical command among mujahideen through the *Layeha*. His simple public persona contrasted with a luxurious private lifestyle, suggesting that his power derived from local perceptions that he was unencumbered by material desires.
(3) *Self-confidence* is "one's sense of self-importance," with leaders high in this trait "more immune to incoming information from the environment" and "not subject to the whims of contextual contingencies" (M. Hermann 2003, 192–194). Taliban commanders have described Mullah Omar as lacking self-confidence before becoming Commander of the Faithful, but he responded to real-time contingencies in abandoning dream interpretation after the United States invaded Afghanistan.

(4) *Conceptual complexity* is "the degree of differentiation that an individual shows in describing or discussing other people, places, policies, ideas, or things," with leaders high in this trait favoring flexibility over consistency (M. Hermann 2003, 195–197). Mullah Omar displayed a capacity for conceptual complexity. His desire to negotiate with rival Ahmed Shah Massoud, proposal to try bin Laden in a country with an Islamic legal system, and admission that the Taliban may have made mistakes after 9/11 denote flexible thinking. However, his insistence on military jihad to expel foreigners was inflexibly rigid.
(5) *Motivation toward seeking office* exists on a continuum, with one end focused on tasks, the other end focused on building relationships, and the middle focused on tasks or relationships depending on context (M. Hermann 2003, 197–198). Mullah Omar fell in the middle. His motivation for office was the implementation of Islamic law throughout Afghanistan, whether in statements from 1998 or statements from 2013. He incited Taliban mujahideen toward this task through speeches praising camaraderie and valor.
(6) *In-group bias* is "a view of the world in which one's own group holds center stage," with leaders high in this trait maintaining the separate identity of the group against a perceived external enemy (M. Hermann 2003b, 199–202). Mullah Omar exhibited high in-group bias in speeches: "By virtue of being Muslim and Afghan, your responsibility is to gird yourself by using your every ability and preparation against the conspiracies of the enemy and to deal with the failed ruses of your historical, cultural, and religious enemy" (Islamic Emirate of Afghanistan 2012a). This identity oriented his diplomacy: "We desire good relations with all those people who respect Afghanistan from the position of a self-determining Islamic country. And their relations are free from barbaric, authoritative, and colonizing tones" (Islamic Emirate of Afghanistan 2012f).
(7) *Distrust of others* relates to *in-group bias* and implies "doubt, uneasiness, misgiving, and wariness about others." Mullah Omar exuded distrust of others, describing the United States as "the greatest evil" and "enemy of Islam." The theme of American and European media as corrupt recurs in speeches: "It has been proven that all of the enemy's evil ruses have been unsuccessful and the secret agencies of the so-called 'free' media has been revealed. And these media people have lost the support of Afghans and

people throughout the world" (Islamic Emirate of Afghanistan 2012d). The corrupt media knowingly conceals NATO losses: "You should know that your innumerable soldiers have been killed in our country and many are becoming the victims of disabilities, and some have become afflicted with mental illnesses due to military pressure. But your governments are hiding these bitter truths from you and your media" (Islamic Emirate of Afghanistan 2013c).

Cultural: A psychological assessment would remain incomplete without cultural analysis. Cultural analysis considers whether the individual and the cultural group would view certain behaviors as normal or pathological and individually or socially determined (Fabrega 1994). This approach can distinguish psychological from cultural elements of leadership.

The Commander of the Faithful as social role:[7] In 1996, Mullah Omar claimed this title after donning a cloak that many Afghans believe belonged to the Prophet Muhammad (Rashid 2002). It is believed that the cloak was worn by the prophet Enoch and presented to Muhammad after his night journey to Heaven (Wood 2009). No Afghan ruler had adopted the title since 1834, when Dost Muhammad declared jihad against Sikh rule in Peshawar (Rasanayagam 2003). Many Afghans attribute mystical powers to the cloak, and rulers have used it to command popular support (Sieff 2012). Sunni Muslim tradition holds that after Muhammad's death (570–632), his successor, Abu Bakr (573–634), assumed the title for continuity of leadership, though the nature and scope of his responsibilities remained uncertain (Kennedy 2004). Another tradition holds that the second caliph (*khalīfa*), Umar (579–644), rejected the title of Successor to the Successor of the Prophet of Allah (*khalīfa khalīfa rasūl Allah*) and encouraged people to call him simply Commander of the Faithful (Crone and Hinds 1986). Besides tribe leadership, the only power the Arabs of Umar's time acknowledged was military alliances, and the new title reflected the military ambitions of the Muslim community (Hodgson 1977). Umar instructed six leaders of the Quraysh tribe to form a council (*shūra*) and select his successor, an institution that has since endured (Donner 2010). By the ninth century, caliphs undertook political leadership, and religious scholars were responsible for theological interpretation (Lapidus 1996).

This cultural foundation clarifies Mullah Omar's leadership. He has drawn repeatedly on the figures of Muhammad and Umar for symbolic power. Muhammad becomes the moral exemplar for mujahideen in Mullah Omar's speeches: "You should behave with love, nobility, and kindness with people in light of the highest morals of the Holy Prophet (May God bestow peace upon him)" (Islamic Emirate of Afghanistan 2012c). Interviews with a former driver confirm that Mullah Omar modeled his leadership after Umar, who frequently met with subjects to discuss problems (Johnson and Thomas 2002). Umar is also valorized extensively: the Taliban's 2012 military campaign Operation Al-Farooq was named after him (Islamic Emirate of Afghanistan 2012e), and the 2013 military campaign Operation Khalid bin Walid was named after Umar's military commander (Islamic Emirate of Afghanistan 2013d).[8]

Mullah Omar's belief in his ability to control events reflects a socially expected role to protect the Muslim community as Commander of the Faithful. However, he seems to have taken a culturally incongruent step in disputing clerics over Osama bin Laden, stressing a need for power and influence outside role expectations. Clerics, not the Commander of the Faithful, have been responsible for theological interpretation, and these scholars could have made the final call on bin Laden. Moreover, Mullah Omar occasionally provided theological interpretations in his Urdu statements, indicating the use of canonical religious texts such as the Quran to frame everyday life. In the eleventh issue of *Sharī'at*, he quotes the twenty-first verse from chapter *Al-Ahzab* (the Confederates) to promote military resistance: "You have had a good example in God's Messenger for whosoever hopes for God and the Last Day, and remembers God oft" (Arberry [1955] 1996, 123). The statement quotes Quranic text in Arabic, followed by this interpretation: "For you people, meaning for those people who fear God and the Last Day and who remember God often, the Messenger of God (peace and prayers of God upon him) was present as a supportive example (meaning in joining jihad) (Islamic Emirate of Afghanistan 2013b, 11). The interpretation departs from the Arabic original in emphasizing fear, and the parenthetical phrase "in joining jihad" refers to Muhammad, two critical phrases absent in the Quranic original. Through these interpretations, an account of Muhammad's conquests fourteen centuries ago becomes a vision for contemporary jihad.

Other leadership traits can be explained through dynamics in Afghan society. In one tale, Pashtun king Ahmad Shah Durrani stole Muhammad's cloak from caretakers in Bukhara (today's Uzbekistan) and installed it in Kandahar in the late 1700s, where it has since remained (Inskeep 2002). An alternative tale holds that the khan of Bukhara gifted the cloak to Ahmad Shah in deference to growing Pashtun influence in Central Asia (Williams 2012). Historians believe that Ahmad Shah seized the cloak from Naqshabandi Sufis in Badakhshan (Gommans 1999). Many Afghans trace Afghanistan's modern history to Ahmad Shah wearing the cloak in 1747 as a symbol of ascendance over competitors after Pashtun tribal chieftains elected him king in Kandahar (Braithwaite 2011).

Notably, Mullah Omar belonged to the Ghilzai tribe, whereas other Taliban leaders belong to the Durrani (E. Rubin 2006). Mullah Omar had to balance power between Ghilzai clerics in Kabul and Durrani clerics in Kandahar (Bernbeck 2010). The similarity between Ahmad Shah Durrani and Mullah Omar is striking: both men wore a cloak popularly believed to have belonged to the Prophet Muhammad in an act of charisma to exude power over rival tribes in the Pashtun heartland of Kandahar. This competition may explain his *in-group bias* toward a pan-Islamic Afghan identity to consolidate power rather than emphasize tribal or sectarian differences.

Omar as xenophobic mullah: Finally, his distrust of others may relate to the cultural role of mullahs in Afghan society. Since mosques receive worshippers daily for prayers, mullahs have served as mosque custodians, congregation leaders, and tribal negotiators with control over male village life (Haroon 2007). Mullahs have joined Sufi holy men and judges in proclaiming jihad following the British invasions of Afghanistan in 1839 and 1878, the fall of the Ottoman Empire after World War I, and the Soviet invasion of Afghanistan in 1979 (Barfield 2005; S. Nawid 1997). Charismatic leadership in Afghanistan before the Soviet invasion traditionally fell to religious scholars and judges with scriptural authority, Sufi saints and tribal leaders claiming descent from Muhammad, and mendicants who rejected the material world for mysticism (Edwards 1986). Unlike the religious scholar, Islamic judge, Sufi saint, or tribal leader whose positions are legitimized by others, the title of mullah is self-conferred, implying someone who has attained some degree of religious education (Kleiner 2000). Omar's status as village mullah in Sangesar endowed him with a power base to attack corrupt mujahideen in a jihad against enemies among religious students. The embedded nature of mullahs within village social structures may lead to distrust of those outside of this system.

This social dynamic corresponds with the Taliban's view of mullahs as leaders. An Urdu author named Painda Khel writes that Mullah Omar embodied the poetry of Muhammad Iqbal (1877–1938), regarded by many as the last great classical Urdu poet, who wrote extensively on transnational Muslim identity (Aggarwal 2008). Khel (2012) explains: "In Allama Iqbal's 'Complaint' and 'Response to Complaint,' Satan explains to the infidels the cure for religious zeal among the Afghans: 'This is the cure for the religious zeal of the Afghans / Remove the mullah from the country.' America has not been able to cure this religious zeal despite all of its efforts. Those wishing to see the mullahs out of Afghanistan should understand that mullahs are in the hearts of Afghans. Infidels cannot remove mullahs from Afghanistan or from their hearts" (23). Khel links the symbol of the Afghan mullah in Iqbal's poetry to Mullah Omar's jihad. His inclusion of this couplet indicates that his audience is fluent in Urdu literature and would accept this cultural justification for Mullah Omar's leadership.

To summarize based on the preceding data, Mullah Omar exhibited not only a striking capacity for simplicity in his worldview, dividing the world into *us versus them* and *right versus wrong*, but also a high capacity for cognitive complexity in his personality and leadership styles by creating and maintaining the Taliban's hierarchy, communication, and resources. The combination of a simple political worldview and complex managerial skills may explain why leaders such as Mullah Omar and Osama bin Laden have been valorized within their communities: their absolute, single-minded loyalty to the group's cause is unquestionable, but they also possess the aptitude to build highly sophisticated organizations to accomplish their missions (LoCicero and Sinclair 2008). This aptitude extends to the level of creating a bureaucracy that structures interactions within the organization, defines standard operating procedures, works toward the organizational mission, and reinforces internal culture—especially when mid-level managers can be killed (Shapiro and Siegel 2012). None of the preceding data suggest Mullah Omar's direct involvement in administering this type of organizational bureaucracy. It is possible that our data sources are limited in this area, but it is more likely that the Taliban has evolved into a decentralized organization where his need to interface with others was limited to protect his security. We shall see in chapter 4 that the Taliban has created a literary genre around interviews with Taliban provincial governors who act as mid-level managers and detail their successes in jihad.

* * *

I have contended that some of Mullah Omar's personality traits reflected social roles and exhibited culturally derived understandings and expectations of leadership, not that his leadership traits could be explained through culture without personal agency. Purely cultural explanations would repeat errors from the mid-twentieth-century culture and personality school, which assumed that culture frames personality for all people equally, without accounting for individual agency or intracultural differences (LeVine 2001). Instead, traits of belief in ability to control events, in-group bias, and distrust of others seem to reflect cultural norms. A lack of belief in his ability to control events or religious motivation to seek office would be culturally incongruent based on his title. A psychocultural assessment allows us to discriminate between cultural and psychological contributions to personality and leadership traits. His high need for power and influence among clerics, low self-confidence in his leadership abilities, and moderate cognitive complexity balancing group relationships with the implementation of Islamic law find no cultural analogues and are more likely individual traits.

Discourse analyses of Mullah Omar's orders also help us understand his leadership in the transnational, deterritorialized virtual emirate. First, the Taliban disseminated his statements to the Muslim community during religious occasions in Arabic, Dari, English, Pashto, and Urdu for diverse audiences. Internet publishing enabled him to reach a global audience that could read texts in simultaneity. However, his orders to the mujahideen appeared monthly in the Taliban's Urdu periodical, raising the question of why these exist separately. One answer may come from an author representing the Media and Cultural Commission of the Islamic Emirate, who explains that Dari and Pashto are "Afghan languages" but Urdu is an "Islamic language":

> Since the Urdu language occupies the status of a language rich in content at a regional and international level, and its speakers are united with Afghans through historical, geographical, and cultural ties, it was felt necessary that

> those who speak, understand, and read this extremely productive Islamic language could obtain correct information about harassment at the hands of the barbaric Americans and about the grief-stricken Afghan nation. (*Sharī'at* 2012)

In chapter 1, I asserted that the Taliban uses Urdu as a regional *parochial* language to reach South Asian Muslims who may not understand the *vernacular* Dari and Pashto languages native to Afghanistan. The preceding quote implies that speakers "united with Afghans alongside historical, geographical, and cultural ties" may be a Pakistani audience, especially since the umbrella organizations of the Pakistani Taliban pledge allegiance to Mullah Omar. Urdu is an essential badge of religious identity for nationalist Pakistanis seeking to promulgate a unitary identity on populations with different regional languages (T. Rahman 2006), especially for Pakistanis of the Sunni Deoband sect, from whose schools the Taliban graduated. The Taliban's cultivation of Urdu resembles twentieth-century history in which Afghan nobility read, corresponded with, and invited prominent Indian Muslim intellectuals to come to Afghanistan and create a Muslim utopia (Green 2011). As languages develop literary cultures, writers must follow normative models for grammar, rhetoric, metrics, and canonicity that bind them to a social identity (Pollock 1998). Here, Taliban authors draw on canonical texts—such as poetry from Muhammad Iqbal—to defend its claims of jihad. The virtual emirate endows the Taliban with a technological platform for the former village mullah Omar to become a global mullah in the transnational Islamic Emirate of Afghanistan.

Mullah Omar's statements in the virtual emirate also reminded followers that Taliban rule over Afghanistan is partial, near, but ultimately incomplete. His interpretations of Islamic law orient Taliban conceptions of sovereignty, and all mujahideen must strive for its implementation, promoting a martial collective identity. This military existence competes with the Afghan government's struggle for governance, producing a situation of horizontal partial sovereignties. Finally, cultural distinctions between right and wrong in the virtual emirate are derived from a fusion of Afghan nationalism and a particular rendition of the Sunni Islamic tradition. Mullah Omar's speeches address Afghans through a nationalistic virtue steeped in war. In terms of religious virtue, the Prophet Muhammad is beheld as the moral exemplar for all mujahideen. Quranic verses and military figures from Islam's first century connect the Taliban's war against infidelity to a valiant past. As village mullah and Commander of the Faithful for the worldwide community, Mullah Omar assumed a leadership role to command respect among a local and global audience simultaneously. In chapters 3 and 4, we shall see how Taliban authors distinguish insiders and outsiders, right and wrong, to encourage shared understandings and expectations in the virtual emirate.

CHAPTER 3

Identity in the Virtual Emirate

THE PRECEDING CHAPTERS underscore the Taliban's constructions of understandings and expectations in the virtual emirate through language: reflecting a foundational Muslim identity, the Taliban builds narrower geographic identities when writing in *parochial* Urdu or *vernacular* Dari. In contrast, the Taliban targets "the entire Muslim world" for fund-raising in *cosmopolitan* English:

> This Islamic land is in the tyrannical grip of the combative infidel enemy, posing the greatest threat to the entire Muslim world. As a result, in the light of Islamic sharia, all Muslims everywhere are duty-bound to join the Jihad with money and soul.
>
> [The] Islamic Emirate of Afghanistan[,] which stood up with [the] Muslim Ummah's physical and financial support, are still waging legitimate Jihad single-handedly with mere help from common[,] sincere Islam-loving masses and is in dire need of financial assistance from the Muslim brothers worldwide for its military and non-military expenditures. (Financial Commission 2012)

If the Taliban now considers "the entire Muslim world" insiders, then why did it subjugate Muslim Uzbeks, Tajiks, Turkomen, and Hazaras before 9/11? How has it since tried to enlist former antagonists and new recruits through self-reinvention with new identities? Do these identities vary by language? This chapter investigates in-group and out-group identities through Taliban texts. I review theories from social psychology on group identity and competition to analyze Taliban texts for representations of insiders and outsiders. Similarities and differences among authors illuminate key deliberations within Taliban discourse. Finally, I consider the Taliban's self-presentation from a pre-9/11, Afghan-nationalist group to a post-9/11, pan-Islamist movement in the transnational virtual emirate.

THE PSYCHOLOGY OF SOCIAL IDENTITY

The Taliban appeal offers an example for thinking through the formation of social identities. In psychology, social identity theory (SIT) studies the social categories to which people identify—such as religion, ethnicity, and nationality—that determine group norms for thinking, feeling, and behaving (Hogg, Terry, and White 1995). SIT originated from Henri Tajfel's (1970) work that "in spite of differing economic, cultural, historical, political and psychological backgrounds . . . the *attitudes* of prejudice toward outgroups and the *behavior* of discrimination against outgroups clearly display a set of common characteristics" (96, original emphasis). Tajfel's (1970, 1974) own experiments and those with colleagues (1971) elicited the following conclusions: (1) social norms are expectations of how others behave and how others expect an individual to behave; (2) in-groups are those to which "we" belong, and out-groups are those to which "they" belong; (3) individuals favor in-groups and discriminate against out-groups; and (4) individuals behave through group norms based on their socialization into the group. In the fund-raising text, the Taliban forms an in-group based on Islam with social norms to support jihad against the "infidel enemy" out-group.

John Turner expanded SIT through self-categorization theory (SCT) by specifying mechanisms for group identification and comparison. Turner (1975) found that group competition arises only when individuals sharing a common trait compete with others along relevant dimensions. For example, the Taliban has not portrayed the powerless Afghan Jewish and Christian minorities as out-groups because they are not "combative." Instead, the Taliban has competed with NATO forces along the relevant dimension of state sovereignty ("tyrannical grip" over Afghanistan), stereotyping them as "combative infidels." Circumstances can change the dimensions of competition deemed salient, but group categories establish common social norms for thinking, feeling, and behaving (Turner 1982; Turner and Oakes 1986; Oakes, Turner, and Haslam 1991).

SIT and SCT explicate processes for group categorization and comparison, but not how group categories exist. Social representations theory (SRT) fills this void by postulating that each society possesses ideas through which individuals can categorize themselves and others (Moscovici 1988). Ideas disseminate through institutions (places of work, worship, and leisure), belief systems (science, religion, morality, and philosophy), and languages that people internalize upon becoming members of society (Moscovici 1994, 1998). Culture can be seen as the process by which individuals interpret the world commonly through shared social representations (Howarth 2002; Moscovici and Marková 1998; Hogg and Reid 2006). Social representations also contribute to society by circulating in texts, media, the arts, and other communication (Ross 2001; Elcheroth, Doise, and Reicher 2011; Staerklé, Clémence, and Spini 2011). Our Taliban author has classified insiders and outsiders through the social representation of religion—despite other available categories, such as tribe or ethnicity—disseminating this interpretation in the virtual emirate.

Although many of these theories arose from experiments in laboratories, they have increasingly been applied to real-world problems. For example, political psychologists have extended SIT to trace the formation of political identities, honing in on ethnic nationalism and conflict (Oakes 2002). Ethnic nationalism can be considered a type of extreme in-group bias against out-group stereotypes based on polarized social representations of a common history and descent (Druckman 1994; McCauley 2001; Finlayson 1998). A cross-cultural approach to studying ethnic conflicts accepts that categories for in-groups and out-groups arise locally and must be understood from an internal perspective to increase the likelihood of mutual understanding and peace (Reicher 2004; Ross 1995). Empirical research on the formation of political identities abandons preconceived categories in favor of social representations from group discourse itself (Moscovici and Marková 1998; Efferson, Lalive, and Fehr 2008), corresponding with the aims of this book.

A DISCOURSE ANALYSIS OF TALIBAN IN-GROUP AND OUT-GROUP IDENTITIES

I have not found scholarship on the Taliban's constructions of in-group and out-group identities. Political psychologist Leonie Huddy (2001) suggests an assessment of identities through (1) the strength of group membership based on who identifies with a group, (2) the characteristics of prototypical group members who embody social norms, (3) core values transmitted to and demanded of all group members, and (4) portrayals of out-groups to clarify intergroup differences. I apply this framework to Taliban texts, analyzing identities by language.

The Strength of Group Membership

English authors employ a transnational understanding of Islam to bolster in-group membership. English author Mustafa Ahmad (2013) incites Muslims in "the West" to rebellion: "The Mujahid brothers abroad should especially target those Crusader soldiers and politicians who have had their hands wet with Muslim blood. The quicker we do these attacks, the quicker the scale of this war will tilt inshAllah [God willing]" (31). Jaffer Hussain (2014) also provokes Muslims in "the West" to sacrifice themselves through "counter-drone strategy": "The brothers in the battlefield of Afghanistan and Iraq have given huge sacrifices but they can't go to New York to take revenge for what the Crusaders and Zionists did in Jerusalem, Fallujah and Kandahar. We are looking towards our brothers in the West to act. Destroy their peace, security, economy and lives as they destroy ours!" (9). Both Ahmad and Hussain use "brother" to reinforce kinship through religion and politicized vocabulary to describe Christians as "Crusaders" and Jews as "Zionists." Hussain (2014) lionizes Muslim "brothers" who attack Americans: "Those who have carried out attacks like our brothers Nidal Hassan, Tamerlan and Johar never travelled to Yemen or Pakistan.[1] They never went through the strict airport and immigration checking. They were the pious explosives in the midst of Kuffar [infidels], who exploded on time for the benefit of the whole Ummah" (9). Both authors divide in-groups and out-groups through religion and use English to reach an audience in "the West."

One anonymous author in *Azan* excoriates officials in Muslim-majority countries for repressing protesters demanding the implementation of Islamic law. The out-group here is not the non-Muslim infidel but the Muslim lacking in faith:

> This Ummah, in Egypt and Bangladesh[,] has had thousands of its sons massacred. They were killed due to their demanding that the Law of Allah

> be implemented on the land of Allah. Who were the killers? They were people who spoke the same tongues and had the same skins as those they slaughtered; but their hearts had been depleted of *Iman* [faith]. After being denied the opportunity to try and establish this Deen [religion] through democracy, they went to the streets in protest. ("Editorial" 2014)

This author castigates Muslims "who spoke the same tongues and had the same skins" for not behaving through social norms in demanding "the Law of Allah." This passage, along with the following English editorial in *Azan*, exemplifies the Taliban's self-positioning as a global Islamist movement beyond its origins in Afghanistan:

> From the battlefields of Khurasan, we give glad tidings to the Ummah of near victory. The withdrawal of America from Afghanistan is imminent and the Emarah Islamiyya ["Islamic Emirate" in Arabic] of Afghanistan is now a reality. Under the leadership of the Commander of the believers, Mulla Muhammad Umar (HA)[2] and the Lion of the Ummah, Shaykh Ayman Al-Zawahiri (HA), the Jihadi movement is now well established and can never be uprooted. ("Editorial," *Azan* 1(4), 2013)

Arabic terms such as "Khurasan" for Afghanistan and "Ummah" for the Muslim community worldwide draw on accepted literary tropes to express a geocultural conception of power (Pollock 1998). The author foresees worldwide conquest: "The Jihad is stronger than ever and it is the Muslims now who will inshAllah [God willing] march forth from the lands of Khurasan, Shaam and the Islamic Maghreb (as indeed they are) to the heartlands of Europe and North America" ("Editorial," *Azan* 1(4), 2013). The use of "Khurasan," "Shaam" for the Levant, and "Maghreb" for North Africa recalls the Arabic names of these regions under the international caliphate starting with the Ummayad dynasty (661–750). The importation of Arabic vocabulary into English strengthens in-group membership around religion and social norms of violent jihad.

Dari authors also promote an in-group identity through Islam, but with national rather than international themes. Muhammad Asim (2010) makes frequent references to the nation (*millat*): "Our proud and angry nation, having overcome many cycles full of turbulent ups and downs throughout history, has always borne witness to the decisions of contemptible and colonizing-oriented people in its era" (9). Asim identifies the colonizers as

Alexander the Great, Genghis Khan, the British Empire, the Soviet Union, and the United States, who have violated the sanctuary (*ḥarīm*) of Afghan territory (9). Habibullah Yusufzai (2013) warns Americans from establishing permanent bases in Afghanistan: "If the Americans envision permanent bases in Afghanistan as they have been claiming, then it must be guaranteed that our nation has never been a nation of a slavish nature and will not sit quietly throughout assaults and attacks. We believe in the strength of our young who will continue the jihad until the last occupying soldier departs from all borders of the country" (59). The unit of analysis is the nation, with "Americans" and "Afghans" substituting for "the infidels" and "Muslims" of English texts to connote in-groups and out-groups, respectively.

Dari author Ahmed Faiz (2010) also appeals to in-group members through nationalism. His text uniquely engages Americans through mutual reasoning rather than military action:

> The Afghan national interest is different from American and Western interests. We cannot accomplish the interests of foreigners and demand sacrifices from our own country, class, values, national identity, and history. Foreigners must respect our will, the Afghan nation's preferences, and the freedom of the Afghan nation and political system according to its own wishes. After these eight years of occupation and transgression upon Afghanistan, it has been proven at the very least that the presence of foreigners in Afghanistan is not in the interest of the transgressors and occupiers or in the interest of the Afghan nation and the region. (25)

Faiz (2010) establishes the in-group through Afghan nationalism and the out-group through foreign occupation, with both groups competing for control of "the Afghan nation and political system." In contrast to English authors with global aspirations, Dari authors define in-group identity through nationalism, leaning on the geocultural trope of Afghanistan's anticolonial struggles to strengthen group membership.

Arabic authors adopt a middle position, basing in-group identity on religion inflected in regional terms of the Muslim world. This editorial from *Al-Somood* (2009) uses Afghanistan as a point of departure to comment on the rights of all Muslims:

> After their long wait, Muslims are strengthened from coming closer to realizing their triumph over their historical enemy (the Americans) in

> Afghanistan. This is a precious opportunity for Muslims to revolt against the enemies of God, the Crusaders, who have shed our blood, trampled our holy sites and our lands, and disgraced our honor. It is up to us all, Oh Muslims, not to lose this opportunity by continuing peacefully against them without adopting a revolt against all of the crimes and calamities that they have committed against our rights, the rights of our world, and the rights of all Muslims. ("Al-Jihād Al-Afghānī Wa Masūlīya Al-'ālam Al-Islāmī")

Shihabuddin Ghaznavi (2009) also advertises the Taliban's relevance to the Muslim world: "It is not a local movement but an international movement whose rays have reached the horizons and whose effects have reached Transoxania, the alpine Caucasus mountains, and the Arab peninsula. Upon its shoulders is the leadership of Muslims in the world" (17). Akram Maiwandi (2010) demands that all Muslims unite: "We return to the Islamic world and its responsibility toward the issues of the *ummah*: the Afghanistan issue, the Palestine issue, the issue of Kashmir, Somalia, Iraq, and others among the hottest issues. And affirm that the responsibility upon the shoulders of the Islamic world is large" (3). In contrast to the Taliban's self-description as an Afghan movement in the 1990s, Ghaznavi (2009) and Maiwandi (2010) position the Taliban as the vanguard of violent resistance against infidels in the Muslim world.

A final example comes from Arabic author Abu Sa'id Rāshid (2010), who begins his text with rhymed prose (underlined), emphasizing the aesthetics of his argument:

> *Subhāna man khalaqa-ki wa dahā-ki, akhraja māa-ki wa mar'a-ki, al-ladhī sana'a min turbati-ki abtālan, raddū min-ki al-dhiāb wa aghwālan, wa al-salawāt ala man arsala ileiki al-jahāfil, taharū-ki min adnās al-wathanīya wa barathīn al-shirk*
>
> Praise to the one who created you and expanded you, who brought forth your waters and lands, who created from your soil heroes who repelled the beasts and ghouls. And prayers upon he who sent to you the hordes who purified you from the transgressions of idolatry and the claws of polytheism. (26)

Rāshid's lexicon—"praise," "prayers," "transgressions," "polytheism"—and themes of creation, purity, and idolatry convert Afghanistan into a site of Islamic history. Rhymed prose (*saja'*) has been a part of Arabic literature since pre-Islamic times, represented in religious and secular literary genres as a remnant of the oral society of desert nomads who used rhyme to remember poetry and information (Stewart 1990; Galander 2002). Rāshid's (2010) text thus fuses style and substance.

These examples epitomize the various strategies of Taliban authors to strengthen group membership. While all authors stress a religious in-group identity, they employ different lexicons, themes, and styles based on language. The geocultural spaces imagined in these texts range from the purely nationalist in vernacular Dari to encompass North America and Europe in cosmopolitan English, passing through the neighboring Muslim world refracted through Arabic. Despite these differences, shared understandings and expectations in the virtual emirate are all firmly rooted in a totalitarian image of Islam that distinguishes good from evil.

Identification with a Prototype

A similar linguistic division occurs with texts on prototypical group members. English texts focus on the living, and Urdu texts focus on the dead, whom the Taliban refers to as "martyrs" (*shahīd*). Arabic and Dari texts mostly focus on the dead, though I have found exceptions in each language. I restrict my focus here to living prototypical members and cover texts on the dead in chapter 4's analysis of jihad.

One Arabic text comes from an early *Al-Somood* interview with Jalaluddin Haqqani (b. 1939), the leader of the Haqqani network based in Pakistan. Haqqani is introduced as "one of the prominent leaders of jihad on the battlefield and the southeastern military official for the Islamic Emirate" ("*Al-Somood*: Tuhāwir Batl Al-Jihadīīn" 2006). Haqqani contends that the situation in Afghanistan is a Muslim matter: "Afghanistan is the country of the Afghan people and the property of the Muslim people. Crusader forces attacked it and the sole solution is the exit of the Crusader forces without any conditions and limitations, leaving the country to its people" ("*Al-Somood*: Tuhāwir Batl Al-Jihadīīn" 2006). Haqqani stresses Afghanistan's anticolonial legacy: "Throughout their long history, the Afghan people have forced many occupiers to leave and retreat from their country, and the best proof of this is the fall of the British Empire and the Soviet Empire at the hands of the Afghan mujahideen" ("*Al-Somood*: Tuhāwir Batl Al-Jihadīīn" 2006). Haqqani ends with a plea to readers: "I say to all of the Muslims in the world that due to the resistance of the mujahideen in Afghanistan, Palestine, Iraq, Kashmir, Chechnya, and Somalia, the enemy faces an abominable defeat, and this is a great matter! I say to all Muslims that it is upon you to form a single rank to realize this great matter and to undertake jihad together against our joint enemy" ("*Al-Somood*: Tuhāwir Batl Al-Jihadīīn" 2006). Haqqani typifies the prototypical group member committed to the norm of jihad against foreign infidels.

In another interview with Shihabuddin Ghaznavi (2008c), Haqqani emphasizes loyalty to Mullah Omar to refute claims of competition between his network and the Taliban:

> I am a part of the high council of the Islamic Emirate and the responsibility for jihad is entrusted to me for the provinces of Khost and Paktika. And based on this, claims from international agencies and the biased media have no truth to them. We have pledged allegiance to the Commander of the Faithful Mullah Muhammad Omar "Mujahid," establishing our allegiance to the present moment. And we are devoted to all of its principles and decisions. . . . And I believe that no commander can be found in the fourteenth century [Islamic calendar] like Mullah Muhammad Omar "Mujahid" in encouragement, generosity, bravery, and action. And this Commander is from the grace of God (may He be exalted above us). All of the mujahideen recognize the power of this grace. (14)

Ghaznavi's interview showcases Haqqani's loyalty to jihad and Mullah Omar as valued social norms. Lest the Taliban feature only high-level commanders, an Arabic text from Sa'ad Nimrozi (2011) describes an attack against enemy barracks. This text displays a psychological candor alternating between fear and fortitude:

> A hill arose before us. At that point, I was with myself and with one of the desires that I had in a battle that frightened me the week before this battle, where I pledged to my Lord: "Oh Lord, if I enter another battle where I meet one of your enemies, then I will kill him until my chest is healed and rage leaves my heart."
>
> Then when one of the soldiers came toward me, I unleashed a barrel of my bullets toward him until he fell to the ground, stained with his rotten blood. With the intensity of my anger toward him in that moment, I stood above his head so that I could unleash a torrent of more and more fire until thirty bullets were lodged in his rotten body. (Nimrozi 2011, 13)

Nimrozi (2011) describes his relief afterwards: "By God, I found in that moment a pleasure and delight that I had not found in any luxury of living, not in food, drink, or even in my wedding night. I say that without rhetoric, ornamentation, or bias" (13). Nimrozi embodies the social norm of acting bravely despite fear.

The sole Dari text in my sample comes from Jamal Zaranji (2014), who has written a series of jihadist travelogues (*safarnāma-e jihādī*) in *Haqīqat*. Travel literature constituted a prominent genre from the fifteenth to the eighteenth centuries among the South Asian, Persian-speaking literati, who wrote ethnographically of themselves and others as cross-cultural contact expanded (Alam and Subrahmanyam 2007). Zaranji (2014) begins with his reason for joining the Taliban: "When the Crusader alliance, with America heading its criminal actions, violated the holy sanctuary of the country and war commenced with the mujahideen of the Islamic Emirate, a group of us young Taliban went from the city of Zaranj, the center of the province Nimruz, toward the battlefield. Our course was the city of Kandahar" (25). Zaranji's text manifests a range of emotions, from anger at a thief who stole the group's money, shock at the expensive prices of hotel food, and surprise that stores were open and children were playing in Kandahar during wartime. Zaranji (2014) recounts his first encounter with a dead Talib's body:

> I was very curious to see the body of this martyr since I had never seen a martyred body with my own eyes before.
>
> I went closer and saw a young man who was very tranquil in the coffin, asleep. I said, "It's like he's alive."
>
> One of the mujahideen answered, "Don't you know that martyrs are alive?"
>
> I went closer and touched his happy and sleeping face. A wondrous smell of perfume had overtaken the ambulance, a smell that I had not sensed anywhere until that moment.
>
> One of the brothers said, "This martyr has a wonderful blessing . . ."
>
> I said, "But isn't this smell of perfume yours?"
>
> He said, "No, brother. This is the perfumed smell of martyrdom. This is the smell of Heaven. This smell is not of this world."
>
> A mujahid who had been our teacher said, "It's as if you are surprised, with your glance of wonder toward this martyr and his perfumed body?"
>
> I said, "Yes, teacher. I am." (26)

Zaranji describes the smell as "musk from the perfumed flowers of spring," "a smell acquainted with the martyrs and soldiers of the Islamic Emirate," and "a munificence (*karāmat*) that raises the delight of any believer toward martyrdom" (26). His text emphasizes social norms of travel, adventure, and the mysterious beauty of martyrdom.

Portrayals of living prototypical members in English also reify social norms of wanderlust for jihad. An interview in *Azan* with a Talib named Adnan Rasheed begins this way:

> Convicted by the apostate Government of Pakistan for his Jihad against the criminal Pervez Musharraf, he spent a hefty time behind bars. He was eventually sentenced to death by the establishment of Pakistan. However, Allah Had Ordained for him a different course. He was freed from his detention facility in Bannu, Pakistan[,] courtesy [of] a memorable operation by the Taliban Mujahideen that saw brother Adnan and scores of other Mujahideen freed from the clutches of the oppressors. He now discharges his duty of Jihad full-time for the sake of the Ummah. . . . May Allah make it a source of benefit for everyone. ("Prison Break: An Exclusive Interview with Adnan Rasheed" 2013)

This Taliban author supports insurrection against "apostate" government forces and travel for jihad as social norms. In another text, Brother Dawood from England describes his embrace of jihad and migration: "Brother, it is more beloved to me to be arrested and put in prison than be a free man walking in the streets of Dar-ul-Kufr [country of infidels]. There's no way I'm going back. It's a one[-]way road. It's more beloved to me walking around in the mountains finding the way to Jihad and dying on the road than to go back to where I came from" (Dawood 2013, 81).

Taliban authors exhort Muslims in countries without an Islamic legal system to migrate for jihad, as in this *Azan* text on two Moroccan Germans:

> Brothers Abu Adam and Abu Ibrahim have had a long journey in Jihad and it would not be wrong to count them among of the veterans of the contemporary Jihad. They have travelled far and wide to reach the Muslim lands. They are of the *Muhajireen* (immigrants) of our times—people who left the glittery life of this world for the glory of the Hereafter. The brothers were born and raised in Germany—deep in the heartland of the disbelieving societies. ("An Exclusive Interview with the German Mujahid, Brother Abu Adam" 2013)

The use of the Arabic term *muhajireen* equates the brothers' migrations with Muhammad's migration (*hijra*) from Mecca to Medina in 622 to avoid an assassination attempt. The author implies that prototypical Taliban members emulate Muhammad, the prototypical figure in all of Islam. In a final English example, Ustadh Abu Imran (2014) consciously presents himself as the inheritor of Muhammad's martial legacy, describing a dream convincing him of jihad:

> In the dream, it was probably the time for *Asr* [afternoon] prayer. I was in the Prophet's Masjid [mosque] of Madina. When I enter the Masjid, the congregational prayer is finished, and they are making the final *Salam* [salutation]. All of the people inside are the *Sahaba* [companions of the Prophet]. I begin to look for Prophet Muhammad, but I can't see him. I then ask one of the Sahaba about Prophet Muhammad, after they pray Sunnah [obligatory prayers] and begin to leave. So they take me to an elderly Sahabi [companion] and this is Abdullah bin Masood.[3] They tell me that if I want to ask something, then I should ask him. So I shake hands with him and sit down. Then I ask him that where is Prophet Muhammad? He says that he has gone for Jihad. So I ask him that why didn't these Sahaba go with him? So he replies that they have a working system—half of the Sahaba go first, and the remaining half go for Jihad when they come back. Then I ask that where the Prophet has gone? He tells me, "North of Afghanistan!" Then I ask him that where would he go after waging Jihad there? He says, "Kashmir." I then ask, "After Kashmir?" He replies, "Palestine!" Then I again make Salam [prayer] and come back. And then I awakened. At that time, my body became very light and my mind became fresh and renewed. Then I became fully satisfied regarding my decision and decided that I would be going to Afghanistan for Jihad. (35)

Imran personifies social norms of travel for jihad, connecting contemporary jihad in Afghanistan, Kashmir, and Palestine to historical figures from Islam's first century, such as Muhammad and his companions. Whether in Arabic, Dari, or English, texts on prototypical group members valorize social norms of bravery, dedication to the Taliban, and travel for jihad.

Core Values

Taliban authors transmit two core values: (1) shortcomings of the non-Islamic international legal order, and (2) the need for traditional Islamic values to halt a foreign cultural invasion. As in previous sections, authors write slightly differently by language.

Belief in the deficiencies of the extant international legal order form a major Taliban core value, since the foundations of this order are represented as coming from human rationality, not Islamic law. Instead, Taliban authors advocate for restoration of the international caliphate. English author Muhammad Qasim (2013c) uses the example of Rabi 'ibn 'Amir—a companion of Muhammad who invited King Rustam (d. 636) of the Sassanian Empire (224–651) to accept Islam—in making a larger point about human fallibility:

> As Rabi' (RA)[4] explained, the Khilafah [caliphate] offers real freedom to the masses—freedom from the oppression of man-made laws and false philosophies that are yielded from imperfect intellects. All religions, philosophies, systems or ideologies other than Islam are corrupt and yield tyranny. That is simply because they are not sanctioned by Allah, Who is All-Knowing, All-Wise. The intellect alone is not a sufficient criterion for creating laws for human systems. (11)

Qasim (2013b) attributes the end of the caliphate system to a European conspiracy:

> Nationalism was meant to achieve two objectives. The first objective: **Expelling Turkey from the region**. The second objective: **Replacing the religion of Islam with the religion of Arabism and emptying the region of its religion so that it would appear receptive to any ideology**. So, it is a European concept, but those who executed it were the Crusader occidentalists in the region. (18; original emphases)

Qasim's (2013b) criticism corresponds to the tendency of English authors to criticize nationalism as a European invention against Islam. Elsewhere, Qasim (2013a) explains: "After the fall of the Khilafah in 1924, this 'country' and 'nation state' concept of the Kuffar was brutally enforced upon the Muslims and since then, generation [*sic*] of Muslims have been born into this 'country'-oriented world. The Divinely legislated brotherhood of the Muslims on the basis of faith has been completely destroyed" (24). Qasim (2013a) cites South Asian Muslim scholars on the caliphate system—such as Maulana Muhammad Ali Jauhar (1878–1931), Maulana Yusuf Ludhyanvi (1932–2000), and Shah Waliullah (1703–1762)—to conclude that "the constitution to the secular state today is what the Bible was to Christianity during its ruling centuries" in contrast to "the Law of the Shariah—which is obtained from Divine Sources: The Book of Allah and the Sunnah of His Messenger" (29). Maulana Asim Umar (2014) goes further in describing democracy as evil: "Shaytan [Satan] has beautified this democratic way for them so much that they cannot even think of leaving it—apart from the seekers of truth" (21). In this telling, Satan tantalizes the masses with democracy as forbidden fruit. Qasim (2013a, 2013b) and Umar (2014) mine the Islamic tradition for historical figures, prominent scholars, and theological interpretations to frame an idealized future of living under Islamic law.

Urdu writer Akram Tafshin (2013a) explains Western fears of a religiously based political system differently:

> After coming out of the dark cave of papacy, secular republicanism was the one clear system of government that each Western power adopted according to its will. Having adopted this system, the West placed it on a pedestal after making it so sacred that speaking against it was like speaking against humanity, even though this is a system and a human-made system at that. There could be a thousand defects and evils in it. However, because this is a system preferred by the West, the right to criticize it also is reserved for the West. (25)

Tafshin (2013a) argues that Europeans rebelled against the Catholic Church during the Reformation after witnessing Muslim progress: "Christian people revolted against the papacy since when they looked at the outside world, they saw man progressing daily in the fields of science and technology and new doors of science and art opening and revealing hidden secrets of the universe, especially in big universities established in the Islamic world" (25). Tafshin's argument departs from those of English authors by portraying the separation of church and state in Europe as a necessity to compete with the more advanced Islamic civilization.

In another variation, some authors censure the international system for not defending innocent Afghans, a unifying theme across languages. Arabic writer Nur Allah Badakhshi (2008) blames the United States for violating international laws: "The barbaric American attack against Afghanistan and the occupation of this oppressed country is against all laws of the United States and its international agreements from day one. And the entry of the United Nations in this field and its authorization to America and its allies to undertake the War on Terror—as they call it—complicated matters and worsened circumstances" (40). Dari writer Ilyas Asim (2011) promises jihad against "powers that daily wish to change the path of destiny for our nation or that attack us in opposition to acquired international norms," since "we forbid for ourselves all of those norms and laws that are in violation of not firmly establishing the religion, values, and interests of our nation" (Asim 2011, 28). Abdul Hadi Mujahid (2014), another Dari writer, warns against substituting Islamic ideals for secular ideals in Afghanistan's political system:

> The laws, standards, and criteria that the West has imposed upon Afghanistan are alien to Islam, and the infidel West has established foundations upon secularism, liberalism, humanism, the philosophy of pleasure and profit, and philosophers—Sartre, Spinoza, Machiavelli, and other atheist Western philosophers—who do not have anything at all to do with fortitude, respect, and pleasing God (may He be exalted). (25)

As proof of the shortcomings in the Western political system, an anonymous Urdu author denounces the lack of international outcry over the deaths of innocent Afghans:

> It is certain that the occupiers will close their eyes to this crime as in the past. Nor will the perpetrators of this crime be brought to court or receive any punishment. But the question arises: If anyone says that he is working for democracy, then can a certificate be given to shed the blood of innocent children and elderly? To eliminate houses, towns, villages, gardens, and lush green orchards from the pages of existence? Does the whole world support and protect this work or is there a higher human standard? Do the international institutions of human rights only apply to those living in America and Europe, and does the blood of other people in this world have no importance next to them? If this is not the case, then what is the reason for the prolonged silence of the human rights institution of the United Nations? ("Kyā Yeh Dehshat Gardī Nahīn?" 2012)

Figure 3.1 provides an example of the images included in Taliban texts to emphasize the oppression of innocent Afghans.

In different ways, Badakhshi (2008), Asim (2011), and the anonymous Urdu author ("Kyā Yeh Dehshat Gardī Nahīn?" 2012) fault the international system for not halting the deaths of innocent Afghans. Writing in vernacular

FIGURE 3.1 *Al-Somood*'s depiction of an American soldier. A picture from the last page of *Al-Somood* 4 (39), published in 2009.

Dari, Asim invokes the "religion, values, and interests of our nation." Writing in languages that travel farther than Afghanistan, Badakhshi and the Urdu author focus on human rights violations and the lack of justice through institutions such as the United Nations or international treaties.

For this reason, writers emphasize core values through Islam, the second major theme. The need for an Islamic morality is seen as a counterbalance against foreign cultural invasion. Arabic writer Zubair Saafi (2008) warns: "We have mentioned that the West wants to change the social aspects of Afghan society and Westernize it. The West has numerous plans and diverse ways to accomplish these goals that are very hidden and that the Afghan people cannot perceive" (34). He names satellite channels in Afghanistan and Arab countries as "the most dangerous," since "they open the doors of houses to the entry of unrestrained Western thought"—an "ideological invasion which the enemies have adopted throughout its stages of conflict with Islam and Muslims" (34). Such channels transmit "programs that promote desires," "lusty films," and "programs of violence," as "enemies have continued to work tirelessly to destroy Islamic and Afghan values" (35). Saafi states that "female licentiousness is presented to viewers under the claim of freedom and strengthening the modern woman" and that "most of what these channels transmit are corrupt habits of Western society such as abnormalities in movements, appearances, dress, food, and moral and behavioral abnormalities; social disintegration; and unraveling, familial disintegration" (36).

Elsewhere, Saafi (2009) cautions against Americans changing values in Afghanistan: "The most dangerous thing confronting the Afghan people, especially in Afghanistan, is the change of identity among the Afghan people generally, and the youth in particular, toward the path of converting Afghanistan, changing it toward America, and its attempt to elevate modern Islam or liberal Islam—which they call 'Moderate Islam'—over true Islam in Afghanistan" (47). Here Saafi cites a news report as proof: "The channel *Al-Jazeera* broadcasted a picture on May 4 showing that American soldiers undertook distribution of the Gospels to local residents near the Bagram military base near the Afghan capital Kabul.... It showed some American officers and soldiers in special religious meetings discussing the importance of spreading Christian teachings among the Afghan ranks" (47). Saafi calls Afghans to defend core values: "It is necessary for us to inform the world generally and the Afghan people especially about religion, that it has individual and social dimensions such as beliefs, forms of worship, and customs that establish its Islamic way, affairs, politics, economics, lifestyle, orders, jihad, education, and upbringing" (48). Saafi's texts thus pit Islam against Christianity and Western thought.

Rather than identifying religious or cultural conversion, Urdu texts harvest the Islamic tradition to find historical parallels with contemporary times. Mullah Abdur Rahman Akhund (2012) connects the bravery of the Taliban to Muhammad's Battle of Badr, when "heavenly angels came from the sky in line after line to help" such that "whatever would have taken months took only weeks" (4). Akhund clarifies: "The Battle of Badr and 'Operation Badr' [of the Taliban] have so many surprising similarities! The name is the same and the work is the same. If anything is different, then it is only that the Afghans have taken the place of the Hijazis" (4). Urdu writer Abu Abdullah (2013) cites the Quran on Muslim solidarity: "It is the teaching of the Quran that you should be one, bind relationships strongly, and avoid factionalism. Similarly, there is a teaching that there are two nations in the world, one Muslim and one infidel" (26). Abdullah rejects nationalism in contrast to Dari authors: "The Prophet—peace and prayers upon him—also narrates that there is no scope for nationalism in Islam" (26). A third example of defending values against a cultural invasion comes through a Taliban statement on education in Urdu:

> The Islamic Emirate supports every education and acquisition of education. It is ready to cooperate in all ways for its progress in accordance with Islamic precepts, and education is a path to progress and salvation in this world and the afterlife. But the meaning of this position of the Islamic Emirate is never that anything is permissible and free under the name of education; it will never be permitted in our educational system that Christian propaganda and constitutions will be used to deviate from Islamic society and that our young generation will be called to another religion other than Islam. ("Talīm Aur Tarbīat Se Mutalaq Imārat-e Islāmīya Kā Elāmīya" 2013)

Deviance from an Islamic morality results in cultural decay. Habib Mujahid (2015) contends that AIDS emerged in Afghanistan as a "gift" of the United States: "According to statistics from the Ministry of Health, in the time of the Taliban's rule 13 years ago, there was no incidence of AIDS. With the arrival of foreign forces and people smitten by the West, AIDS came to Afghanistan and its effects started to spread" (7).

The theme of cultural xenophobia continues in Dari texts. An anonymous Dari writer in *Srak* warns of the War on Terror's cultural consequences: "A cultural invasion is usually invisible and unfelt. It is long term and long lasting. It is foundational and deep. It is from all sides. It is spread on purpose with tools and a program. It is widespread and pervasive. It is productive and dangerous" ("Ātesh Dar Khirman-e Farhang" 2013). The author classifies two such invasions in Islamic societies: (1) "the propagation of ideas and concerns that are not Islamic and even anti-Islamic," and (2) "modernism, science, and extreme rationality, seeking the purifying ways of the West and Westernism, without pointing toward the Islamic sciences and knowledge under the pretext that they do not correspond with man's modern sciences" ("Ātesh Dar Khirman-e Farhang" 2013). Core Islamic values require a defense against progressive Western ideas attacking Afghan society. Another anonymous Dari author in *Srak* includes education within this cultural invasion, targeting the American University of

Afghanistan for non-Afghan values, such as "entering the corner of an alley separate from this country":

> Young sons and daughters gathered independently, in the same place, and acted in a way that completely opposed the social norms of this country. Daughters without hejab or chador went about, something that in other parts of the country would be considered out of the ordinary. Inside the classes, American teachers triggered live debates by using modern technology among the students. In many places, it was seen that this university was an example of a country whose plan of development America had traced out decades before. Some hope that this university would inherit the legacy of America's longest and most expensive war in Afghanistan. ("Pūhantūn-e Amrīkāī-e Afghānistān wa Negrānī Dar Maurid-e Marhala Pas Az Khurūj-e Nīrūhaī-e Khārijī" 2013)

A Dari article by Z. N. (2014) lists the vices accompanying foreigners since the War on Terror: "Problems such as consuming alcoholic drinks and pork meat in domestic markets, the secret establishment of gambling in homes, the promotion of promiscuity, pulling youth toward corruption, the spread of no veiling, the insecurity of the family system, and now, the increasing numbers of addicted children who have acquired the odious politics of the West" (10). Z. N. warns of the West's selfish motives: "In accomplishing its desires, the West does not speak for the value of the lives and futures of the Afghan people. The occupants of the White House think with regard to goals important for them. The number of women and children, old and young, who have been set on the path of sacrifice from their ominous policies is not important to them" (11).

However, Dari texts also reiterate positive aspects of Islamic society rather than only railing against non-Islamic influences. Dari writer Abdul Rahman Hakimzad (2010) encourages family life through Islamic values: "The ultimate wish of compassionate and engaged mothers and fathers is for their children to have a template for life in different areas and to be sources of praise from people, such that upon hearing the description of [their] children from others, they feel happiness and fortune" (49). Hakimzad proposes an Islamic upbringing within the neighborhood and mosque: "It is incumbent upon the sensitive father and guardian to keep the future of his children in sight when choosing a place of residence. And on this basis, to select a place of inhabitance that has good neighbors, a mosque, and a library so that his child develops well from the physical and the educational perspectives with the influence of a good environment and favorable place, to bring forth righteousness and good development (49).

Other Dari writers emphasize hospitality. Maulawi Raza Rukhshani (2013) notes that hospitality has been esteemed throughout history: "Hospitality has now been limited to 'official invitations' for our dear ones or relatives, though the Prophet Abraham used to bring veal to those he did not know solely because they casually greeted him" (81). Rukhshani lists the duties of hosts and guests: preparing food as a mark of affection, preparing food as quickly as possible for the guest, not asking the guest if he has already eaten, eating with the guest, not insulting the guest, always being ready to receive a guest, prioritizing good manners, informing the host about a visit, coming at the stated time and place, sitting at any place indicated by the host, avoiding spying on the household, not making difficult demands on the host, not giving importance to the type of food served, praying for the host, and keeping the host's secrets (81–84). Rukhshani recognizes hospitality as a universal value but situates the Taliban's understanding within the Islamic tradition. We shall see in chapter 4 that the Taliban prides itself on hospitality, even to war prisoners, during jihad.

Differences from Out-Groups

The preceding sections have briefly touched on stereotypes of non-Muslim out-groups perceived as competing for political and military power. In other texts, authors explain the reasons for such stereotypes. English author Muhammad Qasim (2013d) writes: "Instead of the core creed of *Wala* and *Bara* (i.e. a Muslim is an ally to his Muslim and an enemy to the Kafir [infidel]), the new education system preached the lie of the philosophers: 'All human beings are equal regardless of religion; friendship and enmity on the basis of religion is a thing of the past'" (39). Muhammad Ali Sher (2014) rues the forgotten heroes of jihad in the syllabi of secular education: "Names like Khalid bin Waleed, Sultan Jalaluddin Khwarizm, Noor-ud-Din Zangi, Salahuddin Ayyubi, Tipu Sultan, Imam Shamil, Umar Mukhtar (May Allah accept their efforts) fail to strike a chord with us today" (34).[5] Instead, Muslim children are seduced by heroes from Western civilization: "Our children grow up learning about Da Vinci, Napoleon, Shakespeare, Einstein, Lincoln and other such people who were—in the reality of things—none but Kuffar who failed to live their life according to how their Creator asked of them. And their destination is undoubtedly Hell" (34). Qasim and Sher argue for complete separation between Muslims and non-Muslims and an identity steeped in jihad rather than secularism.

Taliban authors treat Christians as despised out-groups. To celebrate *Al-Somood*'s fourth year of publication, Arabic author Haafidh Munir (2009) writes: "When the Americans undertook their invasion of the Muslim countries 'Afghanistan and Iraq,' they gathered all of their monumental media weapons and exploited them for this Crusader war" (2). Munir expands on this American-Crusader conflation: "History repeats itself. How not, as this war is between Islam and infidelity. Bush said with all arrogance and force before his occupation of Afghanistan: whoever is not with us is against us. These are the same terms that Pharaoh used when Pharaoh said, 'I only show you what I see and I will not guide you except on the path of morality" (3). Munir explains the differences between NATO and mujahideen forces: "The big difference between us and our enemies is that they are materialistic, working for the sake of their individual interests and their trivial religious interests. Among us and God (may He be exalted), we do not work for our own interests or for the interest of cheap religious gains; we only want to win great reward which will be the reason for our entry into Heaven" (3). War in Afghanistan comes to represent a cosmic war linked to the Crusades of the Middle Ages and to the infidelity of Pharaoh in biblical times.

Dari writer Saeed Badakhshani (2012) also promotes stereotypes against Christians and Jews by pointing to the video *Innocence of Muslims*, which sparked riots in 2012 among Muslims in many countries:

> Lately, a new culture of defamation, insult, and denial has befallen in a more modern and maddening form from [television] stations and news, filmmakers and media, and especially from the tyrants of history and the headstrong enemies of Islam (Jews and Christians), and at the head of them all, Americans, recalcitrant and stubborn, especially in insult and distortion of the message of the Prophet of Islam, Muhammad the Arab, peace be upon him. (Badakhshani 2012, 74)

In another text, Badakhshani (2014) finds common cause with Christians and Shia Muslims only to disparage Jews:

> In a short period of time, the Jews have tortured many divine messengers with their different behaviors and made them martyrs. They martyred the messenger of the Shias. They sprinkled the blood of the Prophet Yahya [John] on top of the Dome of the Rock for the sake of opposing one of the kings based on his marriage; for this reason, they had also sawed the Prophet Zakariya [Zacharias] in two and martyred him. They took steps to kill Prophet ʿĪsa [Jesus] whom God had sent to save them such that He raised him to the heavens. Many times, this criminal and blood-thirsty group wanted to assassinate the most noble of prophets [Muhammad], peace and prayers be upon him, but each time they were faced with failure. (9)

Urdu texts also exhibit anti-Semitism. Abdul Rahman (2013) writes: "Administrative, economic, social, scientific, educational, governmental . . . these Jews have spread their objectives in every field upon widely influential foundations" (36). Rahman draws an analogy from Islamic history to explain contemporary events: "Since the hidden Jewish hand cannot be discerned behind the war in Afghanistan, their children, having organized an unholy plan to expel the Afghans, are putting into action revenge for the expulsion of the Jews from Medina 1425 years ago. The seditious and corrupt Zionists are drowning the raft of the world, having put forward the new, dirty civilization America" (36).

Even though Taliban authors vilify Christians and Jews in all four languages, varying cultural commitments lead authors to demonize additional out-groups, related to each language's geographical reach. In English, for example, Maulana Abdullah Muhammadi (2013) disapproves of martyrdom status for "heretical" Muslims:

> In Islam, the one who fights and is killed for the Word of Allah to be the Highest is a *Shaheed* (martyr). But in our time, the one who is killed fighting for the country is called a *Shaheed* too. Therefore, even the *Rafidhis* (Shia) and the *Qadianis* who are killed fighting for the "country" are called *Shaheed*. I don't know what Shariah it is that allows even a *murtad* (apostate) or a *mushrik* (polytheist) who is killed for nationalism to be called a "*Shaheed*." (72)

Muhammadi contrasts fighting for "the Word of Allah" and "fighting for the 'country,'" another instance of opposition to nationalism in English texts. Muhammadi draws on Sunni polemics to label Shia Muslims as apostates (*rāfidhī*) and Qadiani Muslims as polytheists. The mention of these groups draws attention to Pakistan, where Qadiani Muslims have been prohibited from calling themselves Muslims since the second amendment to the Constitution of Pakistan was adopted on September 17, 1974.[6] Sect founder Mirza Ghulam Ahmad (1835–1908) proclaimed himself the last prophet of the Bible and the Quran (Ahmad 2010). An anonymous English author in *Azan* explains why the Taliban deems Shia Muslims infidels:

> Shia are not in fact Muslims; rather, they are one of the most dangerous of the Kuffar who are waging a continuous, undeclared war against the real Muslims. From Syria to Iraq to Pakistan to their political stronghold (Iran), the Shias have infiltrated into key positions of the secular democratic systems—be it in the parliaments, the Civil Bureaucracies or the Militaries—and they are utilizing these positions to wage war against the Muslim populations. ("Rawalpindi Massacre" 2014)

This author portrays Shia Muslims as allied with "secular democratic systems" through participation in "parliaments," "bureaucracies," and "militaries"—institutions of governance in modern nation-states. Syria, Iraq, and Pakistan symbolize theaters of competition in which in-group Sunni Muslims compete against out-group Shia "infidels." Examples from previous sections hint that these authors may be writing for a cosmopolitan English readership residing in North America and Europe to represent conflicts in the Middle East from a Taliban perspective.

If the Shia are an international competing out-group, Hindus are regional competitors in South Asia. Urdu author Saiful Adil Ahrar (2012a) writes: "The representative of the Islamic countries, the O.I.C. [Organization of Islamic Cooperation], has been terribly unsuccessful at fulfilling its role. In particular, some Islamic countries are virtually enduring atrocities in hopes of a 'magic' to be released from the dark cover of slavery under the pressure of Jews and Hindus" (40). Unlike writers in English, Arabic, or Dari, Ahrar (2012a) writes in parochial Urdu for a Muslim audience in South Asia, dominated numerically by Hindus.

Finally, Dari author Hafiz Muhammad Khorasani (2013) advises Muslims against celebrating the festival Nawruz due to its Zoroastrian origins: "It is worth mentioning that Hanafi scholars have forbidden celebrating the festival Nawruz since Zoroastrians celebrated the festival Nawruz to appease fire worshippers and they regarded Muslims as infidels. The celebration Nawruz was forbidden because of its legacy as a day of the religion of the fire worshippers" (40). Khorasani appeals to a Hadith—"Avoid the enemies of God by avoiding their festivals"—to emphasize that God views people according to the nations that they resemble. By this reasoning, Afghan Muslims celebrating Nawruz would be infidels like the Zoroastrians, who inhabited Afghanistan, refuted the revelation of Islam, yet claim a common, vernacular Dari literary heritage.

Discourse analyses of Taliban texts affirm the working hypothesis that Taliban in-group and out-group identities vary by language. An in-group identity forged through Islam underlies constructions in all languages, but a sense of regionality pervades these texts through tropes of nationalism in vernacular Dari, South Asian regionalism in parochial Urdu, the Muslim world in cosmopolitan Arabic, and "the West" in global English: the Taliban's own texts indicate that while Arabic and English are both cosmopolitan, English travels wider than Arabic. A geocultural identity radiating along the traveling trajectories of these languages therefore accompanies religious identity. Social identity complexity refers to the phenomenon of holding multiple social identities with relative types of salience, such that in-groups can be defined along a range of categories: merged; equal but distinct; and unequal, in which one dominates another (Roccas and Brewer 2002). In each language, religious identity dominates a geographical identity to strengthen group membership. This point manifests most conspicuously in this chapter's section on core values, derived exclusively from the Islamic tradition, as well as this chapter's section on differences from outgroups, as Dari, English, and Urdu authors target non-Sunni Muslims differently living in areas reached by these texts. Religion is the paramount social representation through which shared understandings and expectations as well as culturally determined differences between right and wrong are defined in the virtual emirate.

The Taliban conception of a militant Sunni, exclusivist in-group advances social values of travel for jihad, absolute in-group obedience, and bravery in battle as exemplified through prototypical group members. This interpretation of Islam departs from other interpretations in which Jews; Christians; and even Zoroastrians, Hindus, and Buddhists were variously deemed "People of the Book" to emphasize the Quran's pluralist message

(Asani 2002). Some Muslim intellectuals in South Asia even attempted to translate and interpret non-Muslims through Islamic concepts and vocabularies toward theological universalism (Ernst 2011b). Instead, the Taliban appeals to an impartial, incomplete sense of its rule in the virtual emirate to instigate in-group members toward violence. The Internet becomes a space through which the Taliban articulates clear notions of cultural difference despite connections with other users.

Textual analysis of social identity through group discourse also offers new research directions. Tajfel, Turner, and their associates elaborated SIT and SCT through the *minimal common paradigm*, an experimental method in which "no social interaction takes place between the subjects, nor is there any rational link between economic self-interest and the strategy of in-group favoritism. Thus, these groups are purely cognitive, and can be referred to as *minimal*" (Tajfel and Turner 1979, 39, original emphasis). However, real-world conditions in the War on Terror have pitted the Taliban against NATO forces with ongoing social interaction through protracted conflict. Apart from rational, cognitive links of economic self-interest and political sovereignty, emotional links also exist: each side has framed the War on Terror as a win–lose, zero-sum game to the detriment of the other side. Taliban authors instead operate through a *maximal exclusive paradigm* in heightening differences with perceived out-groups.

Moreover, analyses of in-groups and out-groups in real-world conditions complicate the relationship between language and culture. Moscovici (1988) defines social representations as "ways of world making" (231), referring to an earlier identification as

> a system of values, ideas and practices with a twofold function: first, to establish an order which will enable individuals to orientate themselves in their material and social world and to master it; and secondly to enable communication to take place among members of a community by providing them with a code for social exchange and a code for naming and classifying unambiguously the various aspects of their world and their individual and group history. (Moscovici 1973, xiii).

Studies of social representations among multilingual groups can consider the extent to which social representations are actually sociolinguistic representations. Moscovici (1973) views language as a medium for social exchange without attending to how language frames the exchange. Languages provide differential codes of social exchange. For example, attacks on nationalism in English texts contrast sharply with elegies to nationalism in Dari. An Arabic author may write through the aesthetic code of rhymed prose, while a Dari author uses the genre of the travelogue. The virtual emirate empowers authors to draw on multiple codes and categories in reaching new and disparate linguistic audiences. We do not know if these texts are intended to espouse a single viewpoint or a multiplicity of views on identity. However, we shall see in chapter 4 that a unitary understanding of jihad defines the actions of in-group members against out-groups.

CHAPTER | 4

Jihad in the Virtual Emirate

THE PRECEDING CHAPTERS have alluded to jihad as an organizational mission: in chapter 1 to justify the Taliban, in chapter 2 as a command of Mullah Omar, and in chapter 3 as an in-group norm against infidel outgroups. Consider this passage from the Taliban's English website:

> The basic meaning of jihad is striving or expending effort, but the word is generally used in a technical sense to denote holy war against the infidels. The Qur'an orders Muslims to expend effort for Allah's cause and to fight for Allah's cause. The Holy Quran specifies that a fifth of the booty belongs to Allah and his messenger, to be used for the common good. It contains a promise that those killed in the battle will have their sins forgiven and will be admitted to paradise. Jihad results in two main things in general for those involved: the first one is to succeed in defeating the enemy; to kill them, to hold them captive, or to make them flee out of the Islamic territory and the second one is to be martyred in this holy way of Allah. In both cases there is striving, but striving for what? Striving for the Creed of Allah being imposed everywhere possible; striving for getting the blessing of Allah Almighty; striving for the rewards of Jannah [Heaven] and forgiveness. In all the three possible results, the Mujahid [striver] is the winner. (Ahmadee 2013)

This author proposes an understanding of jihad as "holy war against infidels" and expectations of either "defeating the enemy" or "to be martyred" for "the blessing of Allah Almighty." The author uses the Quran to frame an idealized future of Heaven and forgiveness through a militarized interpretation of jihad. The act of interpretation is not new; religions acquire new interpretations of beliefs and practices through spreading beyond their origins and guiding the everyday experiences of adherents (Geertz 2005; Poole 1986; Ernst 2011a). Muslim theologians have long attempted to reconcile the Quran as the Word of God with ever-changing historical and social circumstances (Saeed 2005), including political conflicts in modern times (Post 2009). What is new is how jihad acquires meanings in post-9/11 Afghanistan.

This chapter analyzes shared understandings and expectations of jihad in a range of Taliban texts. The previous chapters have mustered evidence to indicate that understandings and expectations can differ by language with convergent and divergent themes. Researchers have suggested the need for data from textual sources such as life histories (Crenshaw 2000), expository texts (Winter 2000), and periodicals (Horgan 2008) to trace how militants represent goals and motivations in lived experience (Crenshaw 2004; Victoroff 2005). For me, this diversity of sources also raises the question of whether different genres lead to different understandings of jihad. Although militant Islamists share a belief in Shari'a law as the sole source of political and moral action in an inherently corrupt world (Euben 1995; Monroe and Kreidie 1997; Taylor and Horgan 2001), an exploration into the act of writing about jihad can emphasize the uniqueness of Taliban interpretations in the virtual emirate.

A CULTURAL PSYCHOLOGY OF JIHAD IN TALIBAN DISCOURSE

Empirical research indicates that social and cultural factors, not psychopathology, inspire jihad. Family honor, social bonds, religious duty, and nationalism have been cited as motivations in interviews with Palestinian widows (Sande 1992), interviews with incarcerated Arab militants (Hammack 2010; Berko and Erez 2005; Post, Sprinzak, and Denny 2003; Merari et al. 2010b), biographies of Al Qaeda members (Sageman 2004), by families and friends of Chechen suicide bombers (Speckhard and Akhmedova 2005, 2006), and last testaments of successful suicide bombers (Kruglanski et al. 2009a). While some have speculated that suicide bombers may suffer from psychopathologies such as trauma and depression (Speckhard and Akhmedova 2005, 2006; Merari et al. 2010a), all studies have affirmed that militant religious interpretations motivate militants. This has fueled the hypothesis that militants may subordinate individual success and self-worth to group interests, fusing individual and group identities for organizational missions (Post, Sprinzak, and Denny 2003). Cultural context, then, determines the balance between individual and group identities, ideologies of collective action, goals judged desirable, and acceptable uses of violence (Moghadam 2003, Silke 2003, Post 2005, Kruglanski et al. 2009a).

Extant studies on Taliban interpretations of jihad lack such empirical cultural analysis. Carsten Bockstette (2009) writes: "The jihadists' primary long-term goal is to restore a devout Islamic caliphate by politically uniting all countries with a Muslim majority in an Islamic realm through a monolithic Islamic religious and social movement. The desired end state is the caliphate's rule worldwide. The Taliban limit their focus on Afghanistan and Pakistan" (7–8). Chapter 5 challenges any characterization of the Taliban as monolithic or limited to Afghanistan and Pakistan. Aneela Sultana (2009) writes: "The extremely strict ideology of Taliban's [*sic*] is considered to be emerged from a combination of *Deobandi* radical interpretation of Islam and *Pashtun* tribal code of honor" (11). Similarly, Syed Hussain Shaheed Soherwordi (2011) writes: "They [the Taliban] were a majority Pashtun movement in a country with a very rich multi-ethnic background, and their interpretation of Sharia was deeply influenced by the Pashtun code of conduct called Pashtunwali" (84). No actual data is presented on Taliban justifications of jihad through interpretations of Islam or Pashtun tribal codes, a necessary point since not all Deobandi Muslims or Pashtuns promote violence.

Sheldon Pollock's (1995) work on the origins of literary cultures, a process termed "literarization," supplies a framework for exploring the construction of meanings in textual genres. Pollock (1995) aims "to relate literary-language choice or change and narratives of literary history to their most salient conditions, the acquisition and maintenance of social and political power," to relate "what such cultural communities might have to do with real or potential political communities" (112). The previous chapters have validated that Taliban authors choose languages deliberately to exude sociopolitical power through differential geocultural themes and styles. The act of writing entails "how at certain times and places certain kinds of language come to be deployed in certain new ways, as never before in their histories, for making certain kinds of texts" as "taxonomized realms of speech" (Pollock 1995, 113, 125). Chapter 2's analysis of Mullah Omar's monthly "orders" exemplifies the novel deployment of Urdu to issue statements to Muslims as the Commander of the Faithful in post-9/11 Afghanistan. Pollock's focus is premodern South Asia, where literary origins are difficult to uncover; my focus is the virtual emirate, from which I have downloaded all issues of Arabic, Dari, English, and Urdu periodicals to trace the trajectories of certain kinds of languages and texts.

Chronicles of Casualties by Date

Taliban publications in Arabic, Dari, and Urdu—though not English—feature short texts on casualties by month. Each *Al-Somood* issue closes with a text from this genre or a table of casualties. *Al-Somood*'s first issue introduced "The Statistics of Jihad," which has since served as a literary model:

> The Statistics of Jihad
> *Jumādā Al-Ūlā* 1427 [June 2006]
> Sources of the Islamic Emirate relate 21 incidents of martyrdom operations (*'amalīya istishhādīya*) and 52 raids (*'amalīya iqtihāmīya*) during the past 6 months in 30 out of 31 Afghan provinces as the mujahideen attacked 171 moving targets (military and air convoys) and 235 stationary targets (centers co-stationing American and Crusader forces, administrative centers, and offices connected to the client Karzai administration). Among which follows a summary of the losses during battles waged in this period.
> Enemy losses
> Human losses:
> The American and Crusader forces 113 killed 163 injured
> The client Afghan forces 309 killed 221 injured
> Material losses
> 6 American combat helicopters
> 1 British vertical plane
> 72 American Hummers
> 50 American tanks
> 40 fuel tanks
> 25 pickup trucks
> 17 motorcycles
> 60 heavy arms, 40 light arms
> 14 launchers, 7 RPGs
> 12 mobile phones and 42 wireless equipment

> Losses of the mujahideen
> Among the mujahideen: 52 martyrs 48 injured
> Among the civilians: 375 martyrs 193 injured
> 19 cars
> 2 motorcycles ("Ihsāīyāt Al-Jihād" 2006)

The use of Islamic months emphasizes the cosmic dimension of jihad. The detail of operations in thirty out of thirty-one provinces highlights the Taliban's national presence. The statistic on American and "Crusader" casualties is 113 for the first six months of 2006, higher than the 75 reported by iCasualties (2006), an independent news source regarded as the authoritative record of casualties in Afghanistan (Bigg 2006). Finally, the Taliban classifies Afghan civilians among the mujahideen, allowing the Taliban to speak for all Afghan Muslims. Subsequent Arabic issues title this genre as "Field Reports" or "From the Trenches of Combat."

This genre displays more narrative in Dari and Urdu Taliban texts. Urdu periodical *Sharī'at* introduced "Martyrs of the Nation" from its second issue, charging the "free" press of the Republic of Afghanistan with ignoring atrocities and deaths of innocent Afghans (H. S. Ahmad 2012). Dari periodical *Haqīqat* introduced the genre from its first issue. Since the genre has the same style in both languages, only one Urdu example is translated. Author Hafiz Saeed Ahmad writes: "We will present the number of casualties of nationals from all corners of the country that are periodically published in government and private media" (2012, 27):

> On the first of March, American forces in an area close to the center of district Delaram in province Nimroz martyred the one known as Haji Ghani Khan in his house for unknown reasons.
>
> On the fifth day of March, district police in the area Khwaja Khadid in Chora district of Uruzgan province raided the house of a local man and entered the house on the pretext of a search. They took away cash and other valuable things. The owner of the house resisted these police steps, resulting in his being martyred along with his family. Two of his children were injured during this resistance.
>
> On the ninth of March, the media reported an account of commissioner Abdul Hakim Akhunzada of Tagab district in Kapisa province that in village Ibrahim Khel of the aforementioned district, NATO undertook a bombing, resulting in the martyrdom of three children, four others, and

> two injured. On that day, in the area named Tarili of Chaparhar district of Nangarhar province, there were two local youths martyred and one injured from the bombing of Western barbarians. (27)

Textual elements include a narrative style, days and locations of casualties, and depictions of perpetrators as Taliban enemies. Names of victims are not consistently reported but labeled "local," reinforcing their identity as "martyrs of the nation." The text ends with an account on March 27 and no conclusion, implying that such acts will continue.

Taliban authors deploy Arabic, Dari, and Urdu to make new kinds of texts on civilian casualties. Each author records statistics akin to modern state bureaucracies that exercise biopower over populations by registering data on birth, marriage, illness, and death (Foucault 1990). The genre supports the Taliban's claims to exercise sovereignty that is fully, flatly, and evenly operative over each centimeter of Afghan territory. Martyrdom in Islam (Arabic: *shahāda*) shares the original Greek connotation (*martyrs*) as "witnesses in law," originating from the early Christians tortured by the Romans to confess their faith (Bonner 2006). The Muslim confession of faith—"There is no God but God and Muhammad is His messenger"—is also named the *shahāda*, implying that those killed by enemies are martyrs by virtue of being confessional Muslims, fusing individual with group identity.

Interviews with Taliban Officials

Taliban periodicals in Arabic, Dari, English, and Urdu also showcase interviews with officials. This genre differs from biographies of prototypical in-group members or hagiographies of slain Taliban militants in that interviews cover military strategies for jihad. The earliest interview appears in the third *Al-Somood* issue and is with the "leader of the mujahideen in Uruzgan province." Arabic and Dari interviews adopt the same style, so only one Arabic example is presented:

AL-SOMOOD We first ask our brother leader to introduce himself to us.

THE LEADER ABDUL MUTTALIB Praise be to God, and peace and prayers upon the Prophet of God, his family, and his companions altogether; and afterward—

Your brother Mullah Abdul Muttalib "Jihadi" was born in village Nawa Dervishan located at the center of Uruzgan province, Tarinkot. I was born in the family of Shaikh Maulavi Muhammad Shafeeq, the previous field leader of Uruzgan province. I have turned 33 years old and I now work in the field of jihad as the official responsible for the mujahideen in the aforementioned province.

AL-SOMOOD What is the number of mujahideen fighting under your command against the Crusaders in Uruzgan province?

THE LEADER ABDUL MUTTALIB About 2200 mujahideen.

AL-SOMOOD What is the number of Crusaders in Uruzgan province and where are they stationed?

THE LEADER ABDUL MUTTALIB 3500 hundred forces from Holland and 1400 forces from the United States are found in Uruzgan province. These forces are stationed in Tarinkot, Deh Rawood, and Khas Uruzgan.

AL-SOMOOD There is no doubt that you have undertaken jihadi operations against them, but how many operations have you joined up to now?

THE LEADER ABDUL MUTTALIB We started jihad against Crusader forces during their attacks against Afghanistan. I have participated in many attacks. ("Qāid Al-Mujāhidīn Bi-Muhāfadha Urūzjān Fī Laqā Ma'a *Al-Sumūd*" 2006)

The interview narrows to logistical details:

AL-SOMOOD Where do you treat your wounded?

THE LEADER ABDUL MUTTALIB During the time of the Soviet invasion of Afghanistan, there was a strong front in our province under the leadership of Maulavi Muhammad Shafeeq and I was one of the people of that front during that time. The front had taken to training some mujahideen in medical knowledge and first aid. Then in the time of the Islamic Emirate, those doctors received basic education in advanced medical centers and are now undertaking treatment of our wounded.

AL-SOMOOD What is the number of your mujahideen martyrs and the martyrs of provincial civilians?

THE LEADER ABDUL MUTTALIB During the start of jihad against the Crusaders, the number of our martyrs reached around 180 and the number of injured reached around 300 mujahideen. But in the barbaric American bombing, many civilians were martyred. Their number reached around 1200 individuals and more than 1500 individuals were injured. ("Qāid Al-Mujāhidīn Bi-Muhāfadha Urūzjān Fī Laqā Ma'a *Al-Sumūd*" 2006)

The official's introduction comprises his (always a male) name, birthplace, and responsibilities in the Islamic Emirate of Afghanistan. A report on jihad follows, with the number of mujahideen and their activities. However, the martyrs reported include civilians and Taliban soldiers, unlike in the previous genre.

Urdu interviews focus less on the officials and more on jihad:

SHAR'ĪAT Janāb [an honorific] Muzammil Sāhib! First of all, please inform us of the performance of the mujahideen related to jihadist activities of Helmand province in the past year.

MULLAH MUHAMMAD DAWUD MUZAMMIL Praise God, the lord of the worlds. Prayers upon the leader of the mujahideen Muhammad, his family, and his companions altogether. And afterward—

In the past year, I want to say that in relation to jihadi activities of Helmand province that, praise God, jihadist activities in Helmand province have been very successful and effective in which the enemy has suffered heavy losses and the mujahideen have achieved many victories. I want to present several points to you related to the conditions of the current jihad over the past year in Helmand province.

1—In Helmand the past year, despite all of the enemy's operations, they were unable to attain even the most ordinary victory and they were not able to bring any territory under their control. On the contrary, they had to wash their hands of many centers that they established which came under their control after withstanding heavy losses during different search operations in 2010. ("Sūba-e Helmand Kē Nāib Amīr Mullah Muhammad Dawūd Muzammil Sē Numāinda-e *Sharī'at* Kī Ek Nishast" 2012)

Mullah Muhammad Dawud Muzammil, the deputy governor (*nāib amīr*) for Helmand province, enumerates two more points: places recaptured and types of attacks. The interview then shifts to military strategy:

QUESTION Would you like to give some information to the readers of *Sharī'at* in connection with the jihadist activities and strategy of the mujahideen?

MULLAH MUHAMMAD DAWUD MUZAMMIL Praise God, the mujahideen are now in a much better situation. It is equivalent to not having any human losses. Through preventive measures and the cooperation of the people, the enemy's raids have been completely unsuccessful—which were the reasons for our

> losses—and the enemy has become dejected. I want to say in connection with the strategy of the mujahideen that just as God (may He be exalted) has commanded in His Word (We will show our way to those who struggle for us [text in Arabic]), God (may He be exalted) has given the mujahideen an effective cure for confronting the technology of the enemy. Here, I will present, as an example, that since human losses have been reaching the enemy mostly through landmines, the enemy has started to put tires in front of its tanks for protection from the landmines so that explosions occur on tires rather than tanks. ("Sūba-e Helmand Kē Nāib Amīr Mullah Muhammad Dawūd Muzammil Sē Numāinda-e *Sharī'at* Kī Ek Nishast" 2012)

Interviews with officials strengthen Taliban claims of sovereignty through the Islamic Emirate of Afghanistan and join jihad in Afghanistan to Islamic history. Each interview depicts American and NATO forces as infidel ("Crusader") and barbaric, confronted by a mujahideen poor in equipment but rich in faith. This trope can be seen as drawing on classical *maghāzī* literature on the expeditions and battles of Muhammad to teach early Muslims that God aided the outnumbered forces of the faithful against larger, better-trained enemies (Faizer 2014). An article from the Taliban's 1998 website locates its struggle within Islamic military lore: "The present Taliban Movement is certainly an Islamic movement and its struggle an auspicious Jihaad. This struggle, this war is in the category of Islamic Ghazwaat (Expeditions of Jihaad)" (Islamic Emirate of Afghanistan 1998a). Taliban authors interpret military themes from classical texts to justify jihad in everyday life and fuse individual identity with group mujahideen identity.

Biographies of Dead Militants

In contrast to other genres, Taliban biographies of the dead in Arabic, Dari, and Urdu commemorate only militants, not civilians. Arabic biographies predate others and undergo an evolution. The first example from *Al-Somood*'s fourth issue, titled "Our Heroes," celebrates Brother *Hāfiz* (an honorific for one who has memorized the Quran) Azeemullah:

> In our renowned Kandahar, our hero grew up through the good tidings of God and thrived there. He adhered to and lived a simple life filled with love and truth. But soon, all of that dissipated when the tyrant oppressed the

> hearts of the believers and gave them a taste of torment and stricture. When he looked at his life around him, he did not find anything that pleased the mind or satisfied the heart. He searched and searched, pure of soul. And it did not please him to sit in his home while his brothers were supporting the tyranny and enmity of the Crusaders and their allies. And it did not please him to be living while his society[1] was filled with corruption. So he turned to God, made up his mind, set out, and hoped to find what could deliver him to his Afterlife. Oh good tidings of God! What is sweeter than martyrdom by exchanging life? And we recalled the story of Haritha (may God be pleased with him) when his mother (may God be pleased with her) came to the Prophet of God (peace and prayers upon him) after the martyrdom of her son in Badr and said: "Oh Prophet of God: You knew Haritha's status among me, so if he is in Paradise I will persevere and be content. And if the alternative has happened, I will not pretend." And he said, "You have grieved. And is there only one Paradise? There are many Paradises and he is in the Paradise of Firdaus." And we ask the Highest Firdaus from God for you, Oh good tidings of God, and we count you among the martyrs. ("Abtāl-Nā" 2006)

The author situates the militant's life within Afghanistan and eulogizes his virtue, contrasting his individual purity with society's corruption. The author also paraphrases a Hadith of Muhammad on martyrdom, connecting the classical Battle of Badr against infidels with war in Afghanistan. We learn of Azeemullah's death:

> On that day, he took an axe—and the axe is kept by people of the area in their houses to guard the house and as a protection against the evil of wolves and beasts of prey—and he went to a meeting of the enemy where the worst of the enemies of God was lurking. The head of the Canadian forces was encouraging the people against the mujahideen. He spit out his words and the Canadian army was around him protecting him. As soon as the lion-hero reached the meeting, he went to the bitterest of the enemies of God, concealing his axe until he approached him. Then he struck him a decisive blow with his axe until he killed him. After he met his target, the Canadian soldiers opened fire on him until he attained martyrdom. His soul departed and he won his desire. ("Abtāl-Nā" 2006)

An Arabic couplet from another Hadith ends the biography: "Galloping toward God without provisions, except for the meeting and the act of the life to come" ("Abtāl-Nā" 2006). The invocation of Hadith endows contemporary war with religious virtue.

Taliban authors change this literary structure in *Al-Somood*'s ninth issue. Akram Maiwandi (2007a) features more than one martyr per article and quotes Quranic verses rather than Hadith texts to begin each article. The verses come from chapter *Ahzab* [the Confederates]: "When the believers saw the Confederates they said, 'This is what God and His Messenger promised us, and God and His Messenger have spoken truly.' And it only increased them in faith and surrender [verse 22]. Among the believers are men who were true to their covenant with God; some of them have fulfilled their vow by death, and some are still awaiting, and they have not changed in the least [verse 23]" (Arberry [1955] 1996, 123). One reason for this change may be that shorter Quranic verses rather than longer Hadith narratives free space for more martyr biographies per article. Another reason may be theological interpretation. The author of Azeemullah's biography has written: "And we ask the Highest Firdaus from God for you, Oh good tidings of God, and we count you among the martyrs" ("Abtāl-Nā" 2006). The construction "we ask the Highest Firdaus from God" can be interpreted as doubt about Azeemullah's ultimate fate. Taliban authors risk exposing themselves to charges of intercession by pleading each martyr's case before God, a contravention of Deobandi beliefs against intercession, which may lead to polytheism (Aziz Ahmad 1964; Robinson 1997). Quranic text without supplication solidifies the connection that dead militants are martyrs in Paradise.

The martyr biography achieves final form in *Al-Somood*'s eleventh issue. The title changes from "Our Heroes" to "Our Martyrs, the Heroes." The introduction cites only *Ahzab*'s 23rd verse and displays the pictures of militants. A literary structure consists of the martyr's childhood, education, jihadist activities, Taliban activities, and death. Maiwandi (2007b) applies this formula to militant Mullah Ghulam Nabi, whose biography I translate completely because this structure persists in Arabic, Dari, and Urdu:

> The Martyr Mullah Ghulam Nabi (*Jihad Yār* [an honorific in Dari meaning "Friend of Jihad"]), may God (may He be exalted) have mercy on him.
>
> The leader of the people, the famous mujahid, the hero of the brave, our brother in God, Mullah Ghulam Nabi (*Jihad Yār*) bin Kakar bin Abdul Baqi achieved the highest level of martyrdom, may God (may He be exalted) have mercy on him.

His birth: The martyr (*Jihad Yār*)—may God (may He be exalted) have mercy on him—was born in the Muslim year 1392—1972 Christian [year]—in village Markulak belonging to Arghandab district, Zabul province. It is located in the south of Afghanistan to whose west neighbors Kandahar province and the provinces Ghazni and Paktika to the east. To the south, the state of Pakistan and a district known for belonging to Kandahar; and to the north, Uruzgan province.

His lineage (*nisab*): The martyr (*Jihad Yār*)—may God (may He be exalted) have mercy on him—belonged to a noble home in the heart of the Tokhi tribe. And it is among the famous Pashtun tribes.

His development: Indeed, the martyr—may God (may He be exalted) have mercy on him—grew up in a home with religion, lineage, and youth based on love of faith toward God, the Great One (*Allah Al-'Azīm*). He had a pure disposition and began his intellectual journey in his youth. He was educated among the *shaikhs* of mosques and traveled from one city to another in search of *Shari'a* knowledge. When he reached the prime of his youth, he saw prohibitions growing, honor violated, and sinful, ignorant rulers engrossed in desires. He was perplexed, ponderous, and unable to bear what was happening in the country. When a harbinger (*bashīr*) informed him of the reformist Taliban movement, which had started in Kandahar province, he left his room and book, answered the call, and joined its ranks. He was unable to complete his education. He continued jihad against corruption until he was martyred and met his kind Lord.

His life (*sīrat*): The martyr (*Jihad Yār*)—may God (may He be exalted) have mercy on him—was skinny of body, medium of height, good of disposition, and good of company. Except that he was angry for the truth and his anger intensified on the battlefield. He possessed strong faith and good actions. He was an honest leader, brave mujahid, and modest in perseverance.

His descendants: His descendants are his son 'Ubaidulrahman who is about 7 years old, four sisters, and many among the mujahideen who have vowed to God that they will follow the line which the martyr—may God (may He be exalted) have mercy on him—drew for them.

When the Commander of the Faithful—may God (may He be exalted) protect him—was informed of his martyrdom, his kind Commander appointed his brother Abdul Ghani (*Jihad Yār*) to his place to lead the convoys of jihad that the martyr urged on.

His jihadist appointments: The martyr (*Jihad Yār*)—may God (may He be exalted) have mercy on him—despite the youngness of his age was a great man who was respected, brave, kind, and known for this. Such that the first time he led armed forces in Gardez, the capital of Paktia province, and the leadership of the front line in the battle of the north fell to him, he triumphed in his appointment over the light brigade. Then he was commissioned with leadership of the defensive brigade in Mazar city, the center of Balkh province. God (may He be glorified and majestic) conquered Bamiyan province through his hands. And he—may God (may He be exalted) have mercy on him—excelled in different military positions.

His bearings: Our gentleman (*sayyid*) (*Jihad Yār*)—may God (may He be exalted) have mercy on him—promoted truthfulness in Islam, steadfastness of the heart in battle, and truthfulness of the tongue in speech. He was good in his actions. He was injured three times with serious wounds. God (may He be exalted) treated him and he returned to his actions without mistrust. One did not see delay from him. God (may He be exalted) supplied him with provisions and the means for accomplishing the Hajj [pilgrimage to Mecca] as a test, but he chose jihad and did not go for Hajj.

His martyrdom: In accordance with God's judgment and his power, and surrendering to the command of God and his order, our gentleman (*Jihad Yār*) was martyred facing God with his good deeds on Friday, the 20th of the holy month of Ramadan, 1422*h*[*ijri*] = 11—23—2001*m*[*ilādi*], the Christian calendar. This was an intense battle that broke out under his leadership near the Kandahar airport in the area Takht-e Pol against the American occupiers and their contractors. He and many of his comrades were martyred through the bombing of combat airplanes belonging to the enemies—To God we all return. (Maiwandi 2007b, 19–20)

The new model includes older elements, such as the martyr's birthplace, moral upbringing, and circumstances of death, though honorifics now stud the text with consistent references to the martyr as "a friend of jihad" and exaltations of God. Nabi is portrayed as so dedicated to jihad that he abandons religious studies and refuses a pilgrimage to Mecca for the Taliban. Themes of a pious upbringing and service to Islam have been used throughout history to situate local individuals among canonical saints and scholars in Islamic biographical literatures (Green 2010). Taliban authors appear to draw on this tradition in situating local militants among religious heroes. Recalling that the Taliban began as a militarized movement of students (*taliban*), there are also distinct Afghan themes. The trope of the *talib* as a traveler at the margins of society devoted to religious pursuits rather than material life has been a feature of Pashto literature (Caron 2012). Azeemullah and Ghulam Nabi qualify as itinerant *talibs* forsaking material comforts—the householder's security, the scholar's salary—for the religious pursuit of jihad. Biographies of dead militants, therefore, are certain kinds of texts using pan-Islamic and Afghan literatures in new ways to fuse individual and group identities.

Martyr biographies also include verses of canonical poetry, suggesting that Taliban authors are paying attention to the aesthetics of jihad. Poetry appears in Dari and Urdu biographies, not in Arabic. Qazi Abdul Jalal's (2013) Dari obituary of Maulawi Abdul Qudus features an unattributed line of poetry from poet Khwaja Shamsuddin Muhammad Hafez (c. 1325–1390): "He whose heart is alive in love never dies; he is firmly established in the register of our world forever" (85). Hafez, as Persian literature's most popular poet, wrote against the dogma of judges, clerics, and scholars in favor of wisdom and broadmindedness (Yarshater 2012). Jalal sees Qudus as an heir to this tradition rather than a figure of institutional authority. A couplet from the poet Abu Muhammad Muslihuddin ibn Abdullah Shirazi (c. 1210–1291), better known as Sa'di, ends Jalal's (2013) obituary: "The name of a good man never dies; he is a man whose name is not erased among the good" (86). Sa'di's work was typically the first book studied for its style in the Persian-speaking world, stretching across Central Asia, South Asia, and East Africa from the fifteenth to the eighteenth centuries (Thackston 2008). Jalal's use of poetry suggests that the martyr embodies goodness and that his reputation will endure like the poets of Persian literature.

Poetry appears more frequently in Urdu biographies, and Taliban authors reap multiple Islamic literatures. Sa'di's verse reappears in a biography honoring the tenth anniversary of Mullah Saifur Rahman Mansur: "They entrust that old corpse under the ground / Whom the ground devours such that no bone remains. / The glorious name of Anushirvan the Just is alive. Even if time has elapsed that Anushirvan has not remained. / Oh so-and-so, do good and count the riches of life / Before the sound comes that you no longer remain" (Hikmat 2012b, 21). This author sees no discrepancy in writing an Urdu biography quoting canonical Persian poetry on the immortality of the pre-Islamic Iranian king Anushirvan (c. 501–579). By the ninth century, Anushirvan became a symbol of the just king in Arabic and Persian poetry (Dabashi 2012). Author Abdul Rauf Hikmat (2013b) uses an unattributed couplet from Arabic poet Abu at-Tayyib Ahmad ibn al-Husayn al-Mutanabbi al-Kindi (915–965) to describe militant Maulawi Abdul Hanan Jihadwal's thirst for martyrdom: "If hearts are big, then bodies tire in their desires" (4). Al-Mutanabbi is heralded for synthesizing prior trends in Arabic literature into a singular corpus that served as an inspiration for centuries (Larkin 2008). Hikmat (2013a) uses poetry from the Pashto warrior-poet Khushal Khan Khattak (1613–1689) in a biography of martyr Hāfiz Badruddin Haqqani: "Our small and big have all gone as martyrs in graves. / Back to back, generation after generation, this craft (*hunar*) belongs to us" (8). The choice of Khattak is notable, since he unified Pashtun tribes into a confederacy against the Mughal Empire and imported Persian loanwords, meter, and rhyme into his writings (Kushev 1997). Hikmat imports Khattak's Pashto poetry to valorize Haqqani within an interminable line of martyrs.

Taliban authors can therefore be seen as engaging in practices of intertextuality—inserting the texts of others within their own—common to Muslim literary networks in South and Southeast Asia. Citations from the Quran, Hadith, religious treatises, histories, and other Islamic texts allowed Asian Muslims to create points of shared experience as a universal Muslim community rooted in local identities (Ricci 2012). The shared experience in Taliban biographies is the glory of martyrdom, manifested in diverse sociolinguistic ways. Biographies of the dead included Quranic verses in Arabic, canonical Persian poetry in Dari, and multiple Islamic literatures in Urdu to aestheticize jihad based on conventions of canonicity—what can and cannot be invoked—by language. The militant obituary becomes a new kind of text, forging a sociopolitical community that subordinates individual to group mujahideen identity.

Expository Texts on Jihad

In justifying jihad, Taliban authors write expository texts in all four languages. The earliest example that I found comes from the first issue of *Al-Somood* in which Ahmad Mukhtar (2006) promotes "martyrdom operations" based on asymmetric warfare:

> In the beginning of the invasion, the world's Crusader forces in general and the Afghan people in particular were indoctrinated that resistance to America and its allies was an impossible act since it had means and materiel which could not be resisted. In addition, a country like Afghanistan, poor and delayed in all respects of life such as the economic, the political, etc. cannot resist the Crusader forces in any way (11).

Mukhtar defends martyrdom operations as a military strategy linking conflict in Afghanistan with others in the Muslim world: "Their strategic military use in the Afghan context points to the Afghan resistance using the same path of resistance that has been used in Iraq, Palestine, and Chechnya" (12). Another example comes from Ahmad Bawadi (2009), who justifies jihad as a defensive act after the end of the Ottoman caliphate: "Our account of jihad here is that it is only a defensive jihad, not an offensive jihad. And the current jihad today within the *ummah* has been since the downfall of the Ottoman state until our day today—this is a defensive jihad, not an offensive jihad. And when jihad and the mujahideen are mentioned in our time, what is only meant by jihad is defensive, not offensive jihad" (42–43). Bawadi urges Muslims to adopt defensive jihad: "If the enemy invades a country of Muslims, and among them is a leader currently adhering to God's orders and laws, it is then necessary to follow him and not go out against him until the crisis is contained" (43). Writing in cosmopolitan Arabic, Mukhtar and Bawadi emphasize a transnational Muslim identity to justify jihad in Afghanistan.

English authors focus on theological, not historical, justifications of jihad. One text differentiates "martyrdom operations" from "suicide bombing":

> Now, the term suicide is generally used to refer to the act of taking one's own life due to discontentment, isolation or otherwise. Suicide is generally considered something blameworthy. Islam, however, has shown us that suicide results when a person runs away from the Commands of Allah and when his relationship with Allah deteriorates. (Anwar and Hussaini 2013, 22)

Citing scholars from all four Sunni juridical schools, Anwar and Hussaini (2013) legitimize jihad through Islamic law:

> The Shariah considers it highly rewarding for a person to sacrifice his life for the sake of Allah with the complete surety that the person's attack will result in his death as in the case of the martyrdom operation if there is

> considerable benefit in it for Islam and the Muslims. The martyrdom operation is a modern technique that is similar to the case of the Sahaba [companions of Muhammad] plunging themselves into the enemy ranks with a sure probability of death. The only difference between the two is that in the martyrdom operation, the operative causes his own death while in the case of the incidents mentioned earlier, the operative is killed by the enemy. (24)

Taliban authors deny that life ends with the demise of the body. Elsewhere, Anwar (2013a) refutes disbelievers who doubt the Afterlife:

> Allah created this life as a test to see which of men would be best in deed. Then after death, everyone shall have to account for their deeds. The good ones shall go to Paradise while the wretched ones will be burnt in Hellfire. Many in the modern world today deny the existence of any life after the life of this world. With the storm of atheism and empiricism that has engulfed the academic circles of our age, the people increasingly believe in a notion that the Quran is attributed to the Kuffar of the time of the Prophet Muhammad. (9)

For this reason, Taliban authors persuade Muslims to give up material gains. Abu Salamah Al-Muhajir (2013) lists common excuses to avoid the obligation of jihad: "But I Fear Death," "But my Parents," "But my Children," "But my Wealth," "But my Beautiful Wife," "But I'm Busy in Da'wah (missionizing)," "But I Need to Increase my Iman (faith)," "I will Make Hijrah and Jihad, but Later," refuting them all: "Free yourself from the shackles of this worldly life and physical pleasures. Yearn for freedom from your constricted desires and embrace the expanse of eternity in Jannah. Don't hesitate; rather begin your preparations for Jihad immediately!" (34).

Authors in parochial Urdu draw on geocultural elements such as canonical poetry to recruit for jihad. In one article, Saadullah Balochi (2012) presents Urdu translations of Quranic verses regarding jihad and poetry to augment his argument, declaring that "God (may He be exalted), praise be to Him, has made us the leader of the world and sent us so that Islam is dominant over the entire world" (34). He reproduces text from the poem *Shikwa* (*Complaint*) by the Urdu poet Muhammad Iqbal on Islam's past glory: "We used to recite the *Kalima* in the shadows of swords. / We who lived, lived for the affliction of wars / And we used to die for the greatness of your name."

The Taliban's conception of Urdu as a South Asian Muslim language also leads certain authors to extend claims across the border into Pakistan. A 2013 editorial in *Sharī'at* on an upcoming conference of religious scholars from Pakistan and Afghanistan states: "There is a nefarious attempt (*mazmūm kōshish*) to establish jihad as an uprising, waste the historic victory of the Taliban in this great continuous war for eleven years, and obtain legal support for the puppet government devoid of support from the Afghan Muslim public established as a result of American occupation. We hope that true scholars will abstain from participating in this conference" ("Kābul Kī Mujawaza Ulamā Kānfarans Aur Ulamā-e Haqq Se Muadabāne Apīl" 2013). An open letter from the Islamic Emirate of Afghanistan to religious scholars states: "The Prophet Muhammad, prayers and peace upon him, appointed the religious scholars, the true scholars, his inheritors. God (may He be exalted) awarded them with a high position and rank in Islamic society and their responsibility is greater than anyone else's" ("Kābul Men Munaqad Honē Walī Ulamā Kānfarans Kē Mutalaq Imārat-e Islāmīya Kā Afghānistān, Saudī Arab, Pākistān, Aur Ulamā- e Deoband Kē Nām Khasūsī Khat" 2013). The letter recalls the aid provided by religious scholars to the mujahideen during war with the Soviet Union before concluding:

> Those religious scholars, whether they are from Afghanistan, Saudi Arabia, Pakistan, Hindustan [India], and the Darul Uloom of Deoband, or another large school or country—you are religious scholars. You know very well that it would be a big misbehavior with our mujahideen brothers who, in reality, are your spiritual children to join this conference and cooperate with the defeated Americans. ("Kābul Men Munaqad Honē Walī Ulamā Kānfarans Kē Mutalaq Imārat-e Islāmīya Kā Afghānistān, Saudī Arab, Pākistān, Aur Ulamā-e Deoband Kē Nām Khasūsī Khat" 2013)

Finally, authors in vernacular Dari recall Afghanistan's colonial past to explain the need for jihad. An interview in *Srak* with a Taliban religious scholar named Maulana Qutaiba equates the Russian and American-led NATO occupations of Afghanistan:

> From the perspective of Islamic Sharia, there is no difference between the transgression of occupation of Russian forces in Afghanistan and the transgression of occupation NATO forces under the leadership of America,

> and each of the two transgressions in Islam having been rejected[,] jihad against the transgressors is a mandatory obligation (*farz-e 'ain*); Allah the Highest has ordered Muslims with jihad and with the freedom of Islam, Muslims, and [their] countries. ("Shaikh al-Hadīth wa al-Qurān Maulānā Qutaiba (Khāksār) Dar Musāhaba Bā *Srak*" 2012)

Similarly, a Dari article by Shahid Salahshur (2014) focuses on the need for jihad within Afghanistan: "Restoring the country is the duty of every person in Afghanistan and for the sake of this duty, the one who seeks to love [the country] cannot undertake anything forbidden" (41). Salahshur interrogates the motives of those forbidding jihad: "We do not deny that the nation deserves restoration, but not in exchange for the values of glory, honor, pride, and the name of Muslims. . . . If they were true to their claims, then why during the rule of the Taliban over Afghanistan did they not come for service? Is this not the same country that existed in those days? The answer is that during that time, there was no money; the dollar economy did not exist" (Salahshur 2014, 41).

Taliban authors create expository texts in each language to reinforce bonds between textual and sociopolitical communities. Arabic texts connect jihad in Afghanistan to conflicts across the Muslim world. English texts concentrate on theological justifications to refute media claims that "martyrdom operations" are suicide bombing. Urdu texts appeal to canonical poetry and the status of religious scholars as arguments to persuade coreligionists in Muslim South Asia. Finally, Dari texts equate jihad with nationalist anticolonial resistance. Each language provides the necessary cultural resources to convince individuals of a group militarized identity.

Jihadist Videos

Videos on suicide bombing and martyrdom have become a cultural staple of militant groups. This torrent of multimedia has inspired new methods of analyzing video discourse. Hsinchun Chen (2012) suggests that videos warrant cultural analysis because they "function as cultural screens for multiple enactments, viewings, and interpretations of accepted patterns, themes, and norms (e.g. suicide bombing, martyrdom) while perpetrating the development of shared understandings" (273). Chen has devised a framework for analyzing jihadist videos through variables such as (1) title,

(2) web source, (3) language, (4) video purpose (documentary, suicide attack, message dissemination, training), (5) victim type, and (6) victim nationality. In the following section, I analyze Arabic, Dari, English, and Urdu videos with the most view counts as of this writing.

English. On September 5, 2014, the video "Alemarah 3" (Islamic Emirate of Afghanistan 2014b) had the most views, at 1,560. The video (total time 36:26) begins with an Arabic recitation of the 39th verse from chapter *Anfal* [The Spoils of War] in the Quran: "And fight them until there is no fitnah [sedition] and until the religion[,] all of it, is for Allah. And if they cease—then indeed Allah is Seeing of what they do" (Taliban translation). Masked Taliban fighters with machine guns zoom in and out of the screen. The video is targeted to "the new young generation of Muslims with Islamic and jihadic ideology" (1:20). Videos of President George Bush, President Barack Obama, and Private First Class Robert "Bowe" Bergdahl contrast with pictures of dead children and tortured Afghans (until 2:23). At about minute three, a verse from chapter *Al-Insan* [The Man] of the Quran is recited first in Arabic and then in English with a Canadian accent. Shots of images from American soldiers in the field are contrasted with speeches from American military and civilian leaders as the Canadian narrator claims that "neither the men, or animals, or insects will be able to remain sane and alive" from American bombing (4:56). "This is the work of the so-called protectors of human rights" (4:58). A shot of detainees blindfolded and handcuffed at Guantánamo is shown as the narrator says, "No need to explain, just look at them yourself" (5:39). The narrator announces: "Even Islam says a nonbeliever's prisoner must be treated well though he is fighting an Islamic army" (7:59).

As a counterpoint, a string of videos appears starring a thinned Private Bergdahl in an oversized American military uniform seated against an ornate carpet. He speaks into the camera about his background, deployment to Afghanistan, and captivity with the Taliban. The video pits the compassion of the mujahideen against the barbarity of the Americans despite their claims as defenders of human rights, a visual manifestation of themes discussed in chapter 3. Most often, Bergdahl is shown in split screen, with images of him eating when he mentions that the Taliban has fed him, or images of American leaders when he discusses their duplicities. American veterans are shown asking President Obama not to escalate troops in Afghanistan (18:19). Private Bergdahl warns Americans that their casualties are being underreported and that they would revolt if they knew the real truth (24:30). The speech is interspersed with Afghan civilians, who blame America for the destruction of their homes, and Guantánamo detainees, who recount their torture (27:35). At one point Bergdahl says, "The mujahideen have never deprived me of food and water. . . . Sleep, I get plenty of sleep. That's all I do here," as he laughs (30:07). He narrates experiences of joking with the Taliban, receiving a toothbrush and toothpaste, taking regular showers, shaving and exercising by himself. He acknowledges that although he is chained by the ankles to prevent escape, "that comes with the territory" of being a prisoner, and he has not been tortured. "I've been getting meals as if I am a guest here" (32:11). "They're taking care of me in the way that their God has told them to take care of their prisoners" (32:32). The video concludes with pictures of detainees waterboarded, blindfolded, attacked with dogs, and subject to other torture as an audio clip from President George Bush is looped: "One by one, the terrorists are learning the meaning of American justice." This video confirms chapter 3's point that the Taliban organizes in-group identity around religion in English media. The victims are all Muslims, and the United States is portrayed as the victimizer.

Arabic. On September 5, 2014, the video "Katāib Badr 3" (The Badr Brigades 3) (Islamic Emirate of Afghanistan 2014e) had the most views, at 1,585.[2] The video (total time 66:38) begins with a recitation of the 25th verse from chapter *Ahzab*: "And God sent back those that were unbelievers in their rage, and they attained no good; God spared the believers of fighting. Surely God is All-strong, All-mighty" (Arberry [1955] 1996, 123–124). A subtitle explains a montage of videos of protesters throwing rocks; burning American flags; marching together; and mobbing buildings in Libya, Yemen, Pakistan, Sudan, Egypt, Palestine, the United Kingdom, and Afghanistan: "Muslim demonstrators in different corners of the world defending the Lord of Prophets" (0:59–2:13). A narrator speaks in Modern Standard Arabic, berating the Crusaders for overthrowing "the *minar* [minaret] of your caliphate, the minar of your community, the minar of your unity," with infidel Muslim leaders despite "your" belief in God, the Prophet Muhammad, and the Quran (3:00). The corpse of Ambassador John Christopher Stevens, killed in Libya in 2012, is shown as the narrator extols Muslims from "east and west, north and south," for protesting slights against Islam, such as the *Innocence of Muslims* video (3:19), discussed in chapter 3. The phrase "The evil Burmese burning Burmese Muslims" captions a video of people being beaten and burned alive, with scorched cadavers of women and children (until 4:29). Images of Israeli war planes bombing Gazan buildings and people fleeing are shown, with young girls and veiled women crying and being killed (until 7:29). The video comes to Taliban troops marching in mountains and protesters attacking Bagram airbase to retaliate against burnings of the Quran. Maps of new countries formed since World War II along with images of former president Hamid Karzai and other Muslim leaders are presented to explain the fractiousness of the contemporary Muslim world (8:12).

The remainder of the video features footage of four Afghan "martyrs" planning and executing attacks. The narrative thread is that jihad in Afghanistan is connected to worldwide jihad to defend the Muslim community. Similar to the obituary genre, each is valorized as a "martyr hero" and praised as "the one possessing virtue" and "the lover of the religion." This screenshot moves into footage of each individual smiling, laughing, and praying before pledging jihad and exhorting the worldwide Muslim community to conduct martyrdom operations as a religious duty. In the longest segment, the video documents the day of the attack. Each soon-to-be-deceased is shown smiling, crying, and embracing others. He moves into action by boarding a truck or donning a uniform. The moments before and after the attack are shot through different camera angles, as is the destruction of soldiers, tanks, helicopters, and unmanned aircraft. Captions for "The First Camera," "The Second Camera," and "The Third Camera," with targets circled in red bubbles, curate the viewer experience, and repeated shouts of "*Allah u Akbar*" (God is Great!) are heard. Montages of Taliban commanders planning attacks using wooden models of buildings and cars as well as satellite images of enemy military bases prove their determination to leave nothing to chance. Two of the four segments end with the narrator proudly reading aloud a media report of the attack. This video confirms chapter 3's point that the Taliban organizes in-group identity around religion by using cosmopolitan Arabic to address the Muslim world. The victims in the beginning of the video are Muslims, but the "heroes" promise to kill as many Jews and Christians as possible, assigning a religious identity to enemy soldiers.

Dari. On September 6, 2014, the video "Al-Sādiqīn" (The Sincere) (Islamic Emirate of Afghanistan 2014a) had the most views, at 5,336. Like its English counterpart, this Dari video (total time 28:04) begins with an

Arabic recitation of the 39th verse from chapter *Anfal* in the Quran: "And fight them until there is no fitnah [sedition] and until the religion[,] all of it, is for Allah. And if they cease—then indeed Allah is Seeing of what they do" (Taliban translation). This recitation is followed by a written Dari translation and a narrator who declares the inevitable victory of the mujahideen: "Under the gaze of the Commission of Cultural Affairs of the Islamic Emirate of Afghanistan, the true knowledge of the Islamic Emirate is sufficiently delivered to the people of the world" (1:55). A narrator explains the meaning of *sincere* according to the Quran: "You are differentiated. We distinguish you. Today, the sincere are those who fight knowingly" (2:15). The video shows three examples of the sincere: (1) those seeking to join martyrdom operations, (2) those imprisoned as detainees, and (3) those fighting on the battlefield (2:41). The video describes jihad as "throughout Afghanistan, from north to south, from east to west, from the heart to the limbs, with the cooperation of all people from all nations and all ethnicities, with a single vision that is firm. And the voice of Islam. And the call for a desired Islam, a demanded freedom, and a rescue of the country from the throes of the international barbarians reaches the ears" (3:51). The narrator emphasizes "the awakening and consciousness" of all people throughout the country (4:03) against disbelief, corruption, bribery, and the killing of innocent people by American and Afghan forces (4:57).

The video pans to various mujahideen attacking Afghan police and military vehicles. A song playing in the background has the following refrain: "The heads of the friends of religion are held high, and the heads of the enemies of religion are slung low." In contrast to English or Arabic videos, the Dari video underscores national unity to combat international tyranny. Images of Taliban soldiers fighting and rejoicing contrast with images of bloodied corpses among Afghan government soldiers; Dari news clips announce official casualties across provinces. Various mullahs rallying crowds of Taliban soldiers for war, with footage of the "sacrifice" (*qurbānī*) and "perseverance" (*tahammul*) of Taliban troops marching against "occupiers" (*ishgālgarān*), "foreigners" (*kharijīān*), and the "hired government" (*hukūmat-e mazdūr*) of Hamid Karzai. The Taliban organizes in-group identity around nationalism in vernacular Dari. No mention is made of other conflicts in the Muslim world, as the focus is purely on Afghanistan, framing jihad against foreign occupation to unite the country's ethnicities.

Urdu. On September 7, 2014, the video "Jihad" (Islamic Emirate of Afghanistan 2014d) had the most views, at 3,716. The video (total time 35:25) also begins with an Arabic recitation of the 39th verse from chapter *Anfal* in the Quran. We hear that "The Emirate's jihadi studio disseminates the latest news at the national and international levels about the current war in Afghanistan" in bringing forward "the victories of the mujahideen and the losses of the enemy" to "unveil the Western media's false propaganda" (1:20). The video features written verses 10–13 from chapter *Al-Saff* [The Rank] in the Quran, with Arabic recitation and Urdu translation:

> O believers, shall I direct you to a commerce that shall deliver you from a painful chastisement? You shall believe in God and His Messenger, and struggle in the way of God with your possessions and your selves. That is better for you, did you but know. He will forgive you your sins and admit you into gardens underneath which rivers flow, and to dwelling-places goodly in Gardens of Eden; that is the mighty triumph; and other things you love, help from God and a nigh victory. Give thou good tidings to the believers! (Arberry [1955] 1996, 275) (3:15)

The narrator beckons in Arabic, "Come to jihad, come to jihad, come to jihad," announcing that "jihad is a great divine obligation" and "jihad is the method of self-determination for an oppressed nation" as Taliban soldiers march and fire weapons (4:04). Text on a split screen announces that "jihad is the fragrance of belief and the standard of bravery" and that "jihad is success and happiness in every case, since the result of jihad is either being a warrior (*ghāzī hona*) or martyrdom" (4:31). Similar texts continue, concluding with "jihad is the most important manifestation of the collaborative unity and mutual sympathy of Muslims" as armed Muslim men of different ethnicities and clothing styles embrace one another in a trench (through 5:35). This video makes the case that jihad is obligatory for all Muslims.

The narrator announces that the purpose of the video is message dissemination, "to arouse the jihadi passions of viewers" by depicting the Taliban's victory over six enemy bases in Nuristan province. Montages of mujahideen embracing, marching, firing weapons, and yelling "*Allah u Akbar*" appear as the narrator reminds viewers that Muhammad made his home not among worldly pleasures but among weapons in defending Islam, possessing nine swords and three daggers (8:25). Like written biographies rehearsing the *maghāzī* themes of Muhammad's military expeditions, the video links contemporary martyrdom to Muhammad's sacrifices. The narrator laments that Muslim youth ignore Islam's military heroes as names of historical Muslim warriors flash on the screen. The video includes mujahideen across the ages; Taliban mullahs lecturing on jihad to the mujahideen; and the mujahideen marching across cliffs and mountains to take battlefield positions, with footage from various angles. Visuals are complemented with three types of overlaid audio: shouts of "*Allah u Akbar*" upon the firing of weapons, conversations among mujahideen in Pashto, and various songs in Urdu extolling bravery. The refrain of one song repeats: "We have fulfilled the promise of love, O Lord; We have made Ahmad [another name for Muhammad] our leader before our eyes." American flags are shown lowered at check posts as Taliban militants raise their flag and loot weapons from corpses of "the Crusader occupiers." This song symbolizes the Taliban's vernacularization of Arabic into Urdu. A verse is transcribed and translated here, indicating Urdu and Arabic portions:

> Khudā kē dīn kē līyē, Yeh sarfarōsh[3] chal pare [Urdu]
> (These impassioned ones have embarked for God's religion)
> *Khudā kē dīn kē līyē, Khudā kē dīn kē līyē* [Urdu]
> (For God's religion, for God's religion)
> *Sabīlu-nā sabīlu-nā, Yeh hi hai apnā rāstā* [Arabic, Urdu]
> (This is our path, this alone is our path[4])
> *Sabīlu-nā sabīlu-nā, Yeh hi hai apnā rāstā* [Arabic, Urdu]
> (This is our path, this alone is our path)
> *Sabīlu-nā sabīlu-nā, Al-jihad al-jihad* [Arabic]
> (This is our path, jihad, jihad)
> *Al-jihad al-jihad, Al-jihad al-jihad* [Arabic]
> (Jihad, jihad, jihad, jihad)

The author of this song demonstrates poetic proficiency in Arabic and Urdu, since the two hemistiches are translations. The Urdu translation of the Arabic *Sabīlu-nā sabīlu-nā* is especially noteworthy, since the Arabic means "This [jihad] is our path" but the Urdu *hi* emphasizes that "this [jihad] alone is our path." The composer attends to the aesthetics of the message in crafting Arabic and Urdu lines in the same meter. I scan one fragment through short (-) and long (=) syllables for each of the eight syllables in the lines:

1	2	3	4	5	6	7	8
Sa	bī	lu	nā	sa	bī	lu	nā [Arabic]
-	=	-	=	-	=	-	=
Yeh	hī	hai	ap	nā	rā	s(a)	tā [Urdu]
-	=	-	-	=	=	-	=

The lines do not scan perfectly in the fourth and fifth syllables. However, this would not be concerning to an Urdu audience, since Urdu meters, inherited from Arabic, demonstrate the challenges of fitting the grammar and rhythms of one language within another language's style (Pritchett and Khaliq 2003). The symmetry reinforces the aesthetics of jihad and matches other Taliban chants intended for memorization through poetic balance (Johnson and Waheed 2011). The symmetry also creates a point of shared aesthetic experience, as meters from cosmopolitan Arabic are applied to parochial Urdu. Weapons of looted booty are displayed (until 30:57) as wounded mujahideen are shown convalescing under the care of a healthcare professional with a stethoscope, conducting physical exams. The video concludes with a reminder that jihad is obligatory and a blessing bestowed on the entire Muslim community. This video confirms the Taliban's ideology of the parochial Urdu as an Islamic language, one that travels outside Afghanistan but not diffusely throughout the entire Islamic world (like Arabic) or to North America and Europe (like English).

• • •

This chapter shows that Taliban authors create new genres of texts on jihad and deploy languages in novel ways to endow cultural communities with sociopolitical awareness. Short texts on casualties demonstrate that the status of martyrdom is extended to all Afghan Muslims killed in war compared to obituaries that only commemorate militants. Interviews with Taliban officials and jihadist videos concentrate on military strategies, whereas expository texts justify jihad through arguments that resonate with sociolinguistic themes. In all cases, Taliban authors mine the wealth of the languages in which they write, from invoking canonical poets to forging new literary styles based on classical textual models. The literarization of jihad in Taliban texts establishes that goals and motivations of militants can vary by language and genre, with the need to triangulate data from multiple sources in future work. Despite such differences, each text reiterates the call to subsume individual success and self-worth to group mujahideen identity. Although the meanings assigned to jihad vary, they always justify violence for Taliban authors, whether framed as military strategy in asymmetric warfare, revenge for historical wrongs, or conflict justified by theological interpretation.

Texts on jihad also reiterate themes in the Taliban's virtual emirate. The range of shared understandings and expectations embedded in individual psychology differ by the type of text accessed through websites. The Taliban presents casualties by calendar date and interviews with officials to support its claim of performing law-and-order functions, revealing the patchwork of multiple, horizontal partial sovereignties in Afghanistan. Finally, Taliban authors interpret canonical texts such as the Quran and Hadith to frame an idealized future immersed in jihad. The Internet allows Taliban authors to present themselves as experts in theological interpretation and to position themselves as an alternative source of authority to religious scholars who may otherwise disagree with their interpretations—a crucial political move, since many *talibs* have not yet completed formal religious education. In chapter 5, we shall see how the Taliban extends the notion of jihad to out-groups through a culturally distinct form of foreign policy.

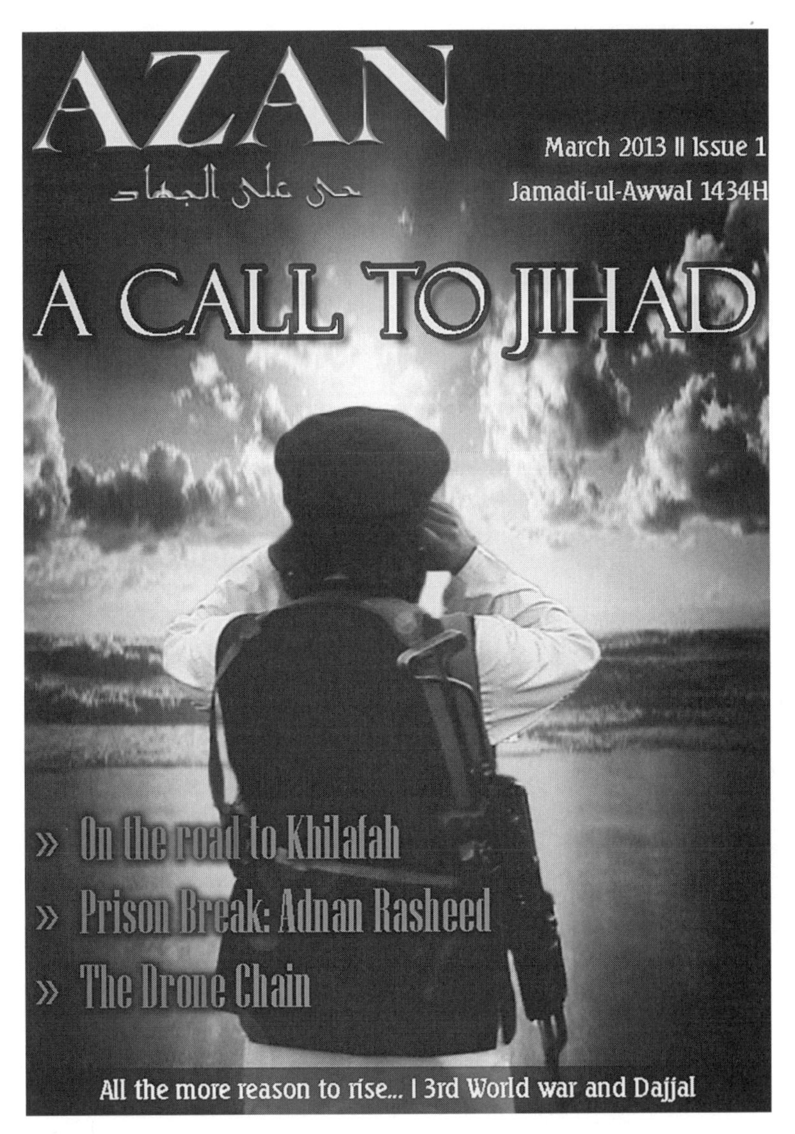

AZAN
حى على الجهاد
March 2013 II Issue 1
Jamadi-ul-Awwal 1434H
A CALL TO JIHAD
» On the road to Khilafah
» Prison Break: Adnan Rasheed
» The Drone Chain
All the more reason to rise... I 3rd World war and Dajjal

The name of the Taliban's English periodical is *Azan*, meaning a "call to prayer" in Arabic. This image depicts a man performing the call to prayer in traditional stance, with both hands at the ears. The *pakol* cap identifies the man as an Afghan, and the gun slung around the right shoulder emphasizes the militaristic aspects of jihad.

Think not of those who are killed in the way of Allah as dead, Nay they
are alive, with their lord, and they have provision.
[Al-imran 3:169]

This image features bilingual text against an idyllic landscape. The stylized Arabic text in the sky reads "*Allahu Akbar*" (God is great). The quotation from the Quran refers to the belief that those killed on the path of God are not dead but martyrs living with God. Quranic quotations are one way that the Taliban justified jihad. The landscape depicts a lush verdant field against a brilliant blue sky, likely depicting heavenly Paradise.

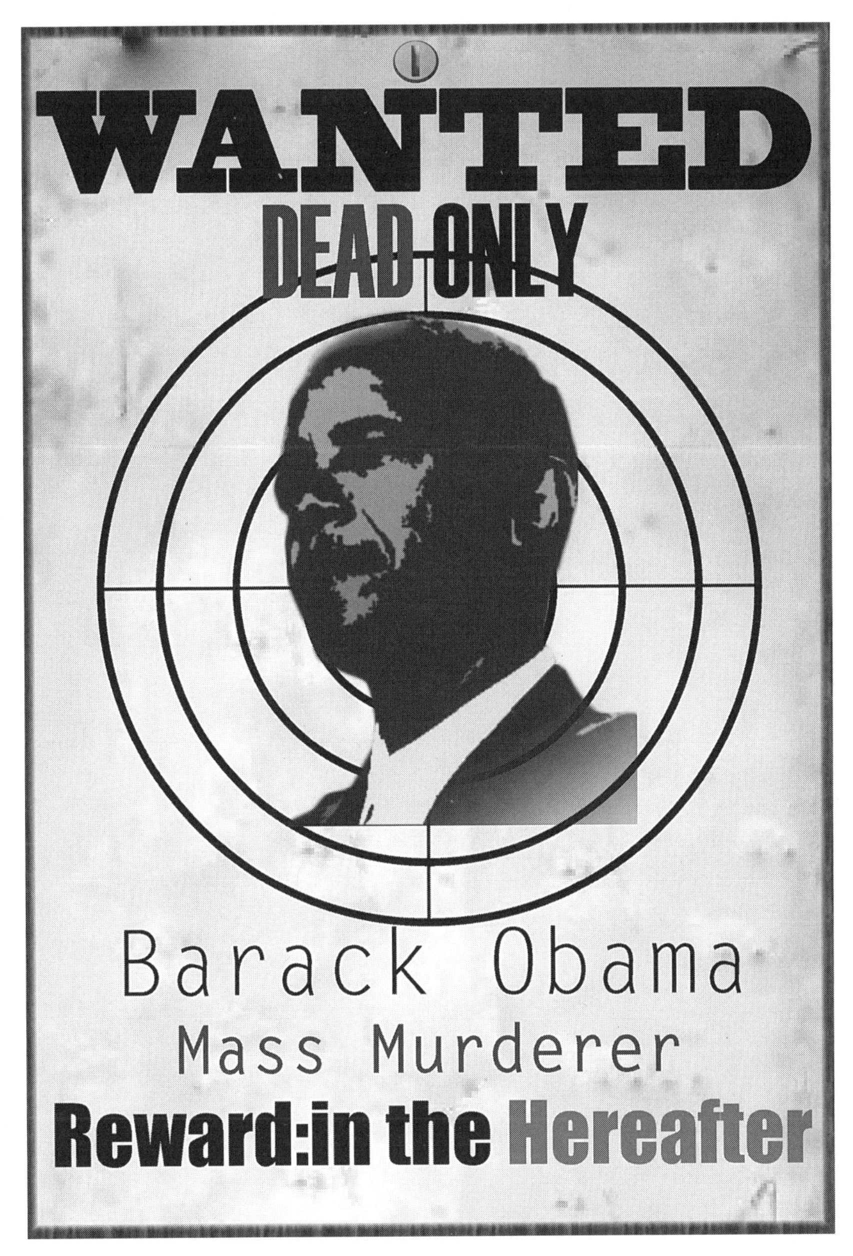

American foreign policy is a frequent topic in Taliban texts. This image features President Barack Obama in crosshairs, a response to the FBI's Most Wanted posters of Mullah Omar, the Taliban's last supreme leader. Taliban texts blame the United States and NATO allies for killing civilians indiscriminately—hence, the epithet "mass murderer." The reward in the hereafter emphasizes religious justifications of violence.

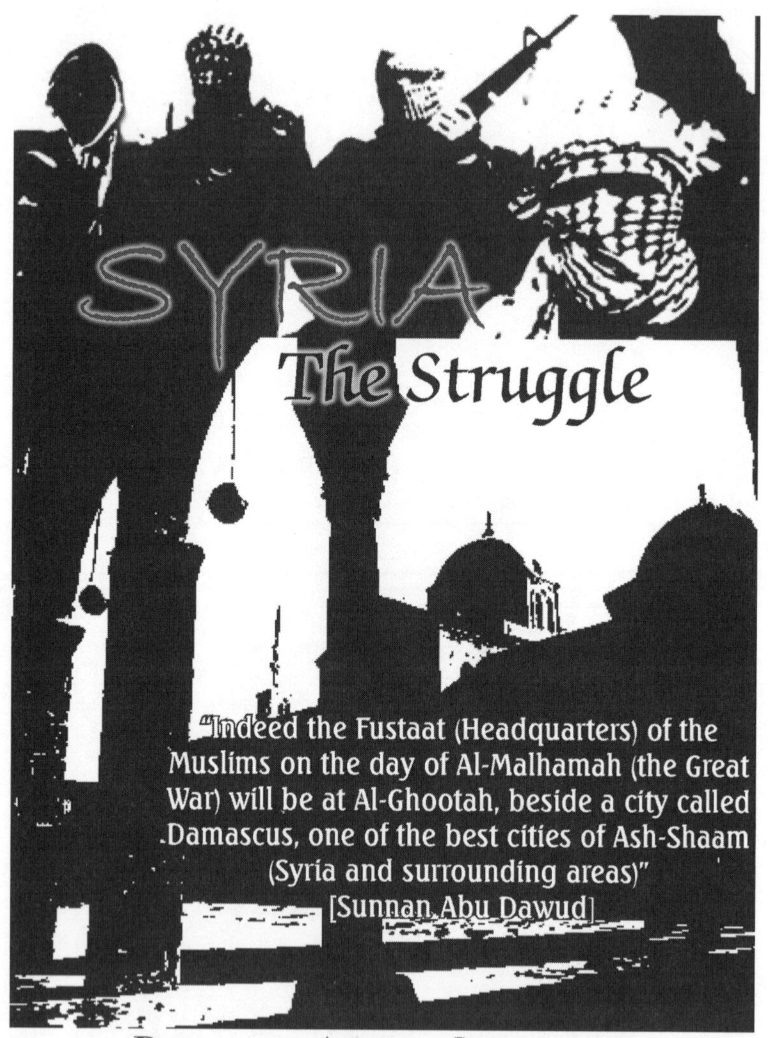

SYRIA
The Struggle
"Indeed the Fustaat (Headquarters) of the Muslims on the day of Al-Malhamah (the Great War) will be at Al-Ghootah, beside a city called Damascus, one of the best cities of Ash-Shaam (Syria and surrounding areas)"
[Sunnan Abu Dawud]
BILAD ASH-SHAAM

The Taliban has increasingly positioned itself as an organization that speaks on behalf of Muslims worldwide. This image depicts militants at top and the Umayyad Mosque in silhouette. Islamic eschatology holds that Jesus will descend at the Umayyad Mosque in Damascus during the End of Days. The quotation comes from the Book of Battles of the chronicler Abu Dawud. The Taliban is attempting to link the contemporary civil war in Syria to the End of Days and valorize jihad.

WE ARE FIGHTING IN
Khurasan BUT OUR
EYES ARE ON
Palestine

This militant stares into the distance at an image of the Dome of the Rock in Jerusalem. Many Muslims believe that Jerusalem is the third holiest city in Islam, after Mecca and Medina, since the Muslim Prophet Muhammad ascended to Heaven from the Dome. The author uses the words "Khurasan" for Central Asia and "Palestine" to link militant struggles worldwide, as the Taliban seeks to build a transnational identity.

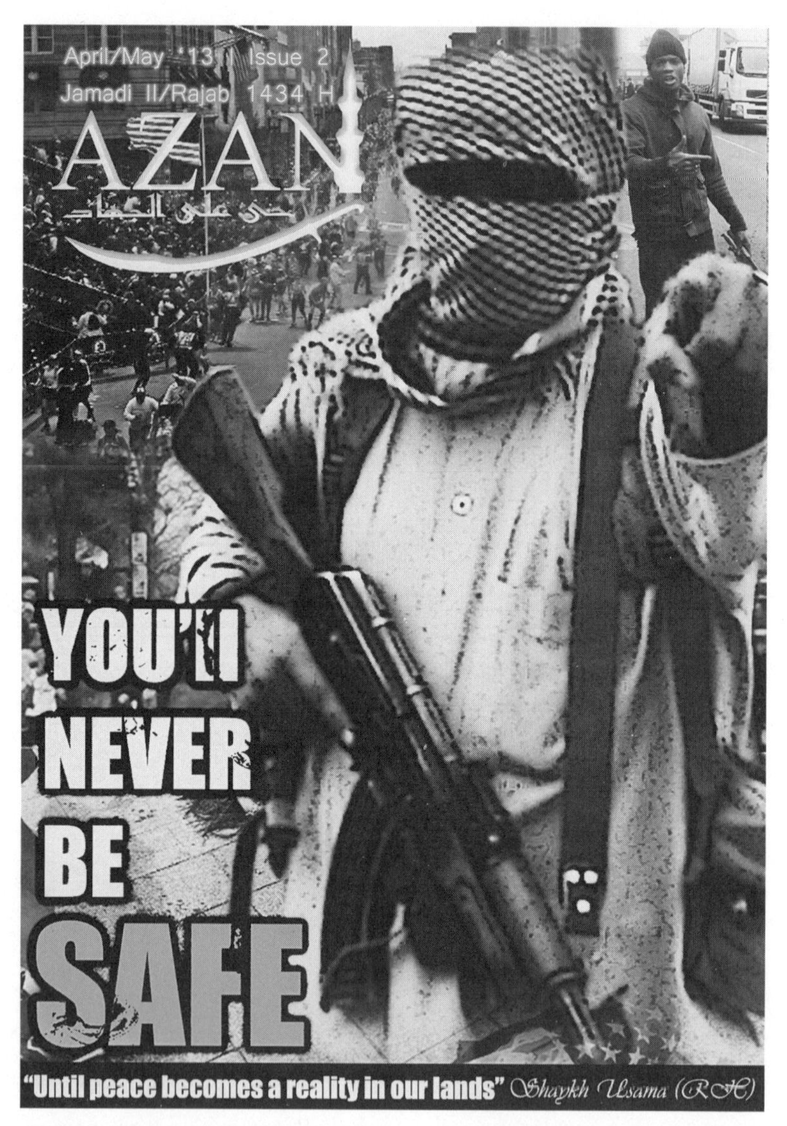

April/May '13 | Issue 2
Jamadi II/Rajab 1434 H
AZAN
YOU'LL NEVER BE SAFE
"Until peace becomes a reality in our lands" Shaykh Usama (RH)

In this image, a militant threatens Americans, identified by the flag in the background. The quote from Shaykh Usama (RA) refers to Osama bin Laden, the deceased leader of Al Qaeda who swore allegiance to Mullah Omar, the Taliban's deceased supreme leader and supposed Commander of the Faithful of the worldwide Muslim community. *RA* is an English acronym of the Arabic expression *rahamahu Allah*, "May God have mercy on him."

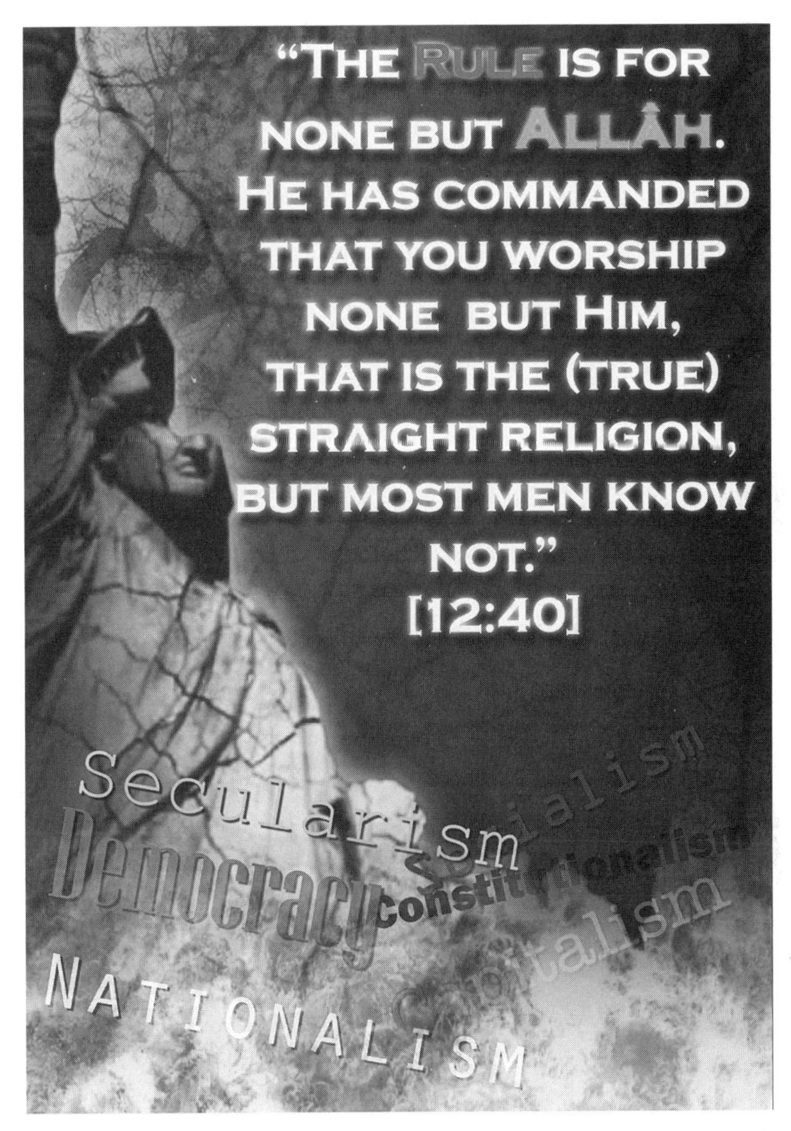

"THE RULE IS FOR NONE BUT ALLÂH.
HE HAS COMMANDED THAT YOU WORSHIP NONE BUT HIM, THAT IS THE (TRUE) STRAIGHT RELIGION, BUT MOST MEN KNOW NOT."
[12:40]
Secularism
Democracy
NATIONALISM

Taliban texts criticize non-Islamic political systems for allowing humans to build laws. In contrast, the Taliban believes in an Islamic system in which laws and legal reasoning are held to be descended from God. Humans can only act as interpreters of ethical, legal, and political precepts that are steeped in Islam. This image portrays the Statue of Liberty and other concepts—such as "secularism," "nationalism," and "democracy"—as false idols to Allah that humans mistakenly worship.

PROPHET MUHAMMAD ﷺ SAID:
"FREE THE PRISONER!"
REMEMBER YOUR IMPRISONED BROTHERS AND SISTERS WITH YOUR PRAYERS, WEALTH AND EFFORTS...
Adnan Rasheed

In this image, a quotation from the Muslim Prophet Muhammad is used to justify freeing those imprisoned for militant jihad. Adnan Rasheed is a Taliban militant who escaped jail in Pakistan. An interview with him in the Taliban's English periodical *Azan* discusses his imprisonment for waging war against the Pakistani state for not implementing Islamic law completely.

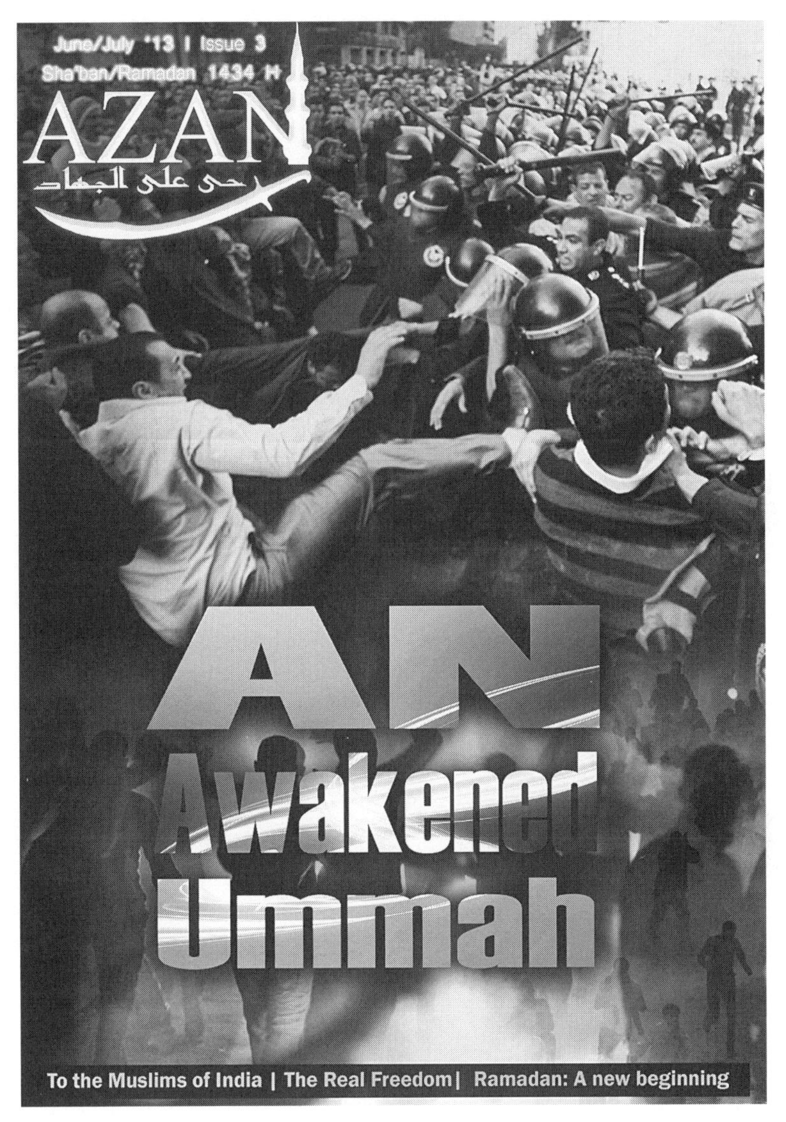

In this image, a mob of protesters battles the police. The phrase "an awakened *ummah*" refers to an awakened Muslim community. The awakening refers to a rising political consciousness that does not assume that the police represent the people's interests. The date of June/July 2013 and the police uniforms suggest that this image refers to a country in the midst of revolution during the Arab Spring.

In this image, the Dome of the Rock in Jerusalem is held behind barbed wire. The expression "47 Ramadan under Jewish control" refers to forty-seven years under which Palestine has been under Israeli occupation. The Taliban constructs insiders and outsiders on the basis of religion, conflating Jewish, Israeli, and Zionist identities.

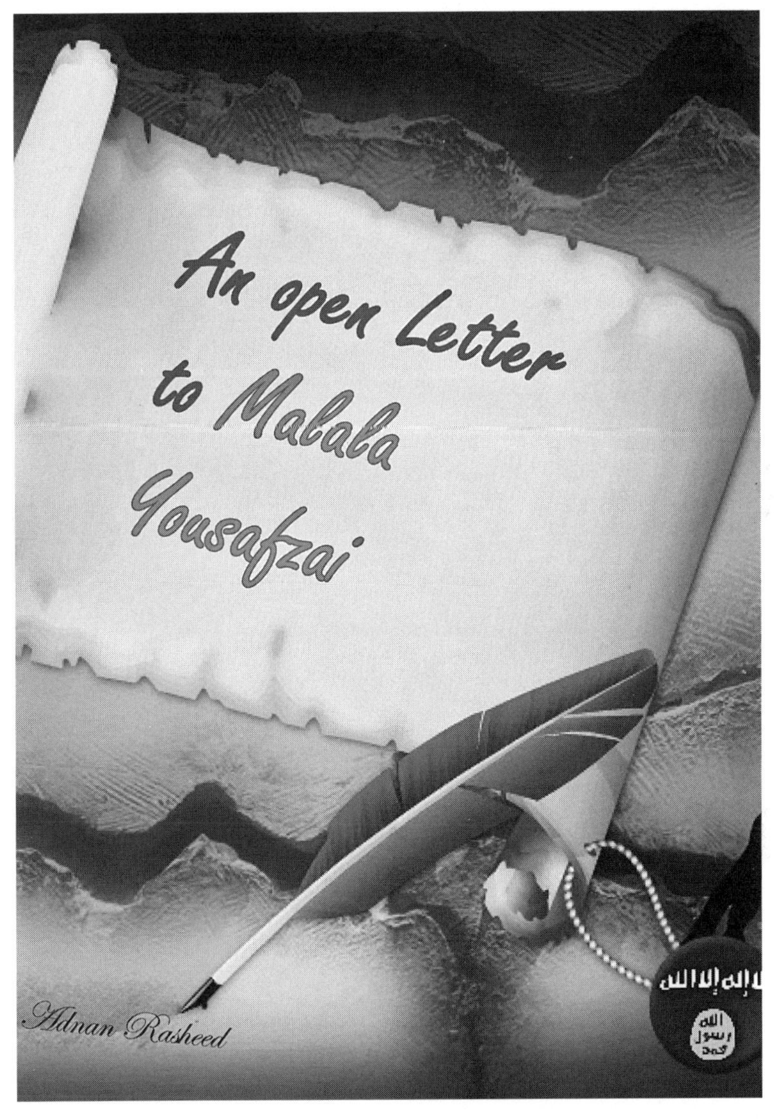
An open Letter
to Malala
Yousafzai
Adnan Rasheed

Taliban authors frequently comment on current events. In this image, Taliban author Adnan Rasheed pens a letter to Malala Yousafzai, the Pakistani Pashtun girl who defiantly attended school despite the Taliban's threats and lived through a subsequent murder attempt. The image evokes a theme in early Islamic history in which the Muslim Prophet Muhammad wrote letters to heads of state demanding that they accept Islam. The Arabic text at the right lower corner is known as the "Seal of Muhammad." The first line comes from the *Shahāda*, the Muslim profession of faith, and is translated as "There is no God but Allah." The bottom three words are "God, Prophet, Muhammad" and are reputed to be in Muhammad's handwriting. The Seal of Muhammad has also been adopted by militant groups such as the Islamic State of Iraq and Syria, also known as ISIS or IS.

In this image, a Quranic quotation is used to justify a militant interpretation of jihad, personified by a man with a gun.

Taliban periodicals openly seek to recruit for jihad through glossy, high-resolution images. In this image, a Quranic quotation justifies jihad and emphasizes bonds of group solidarity.

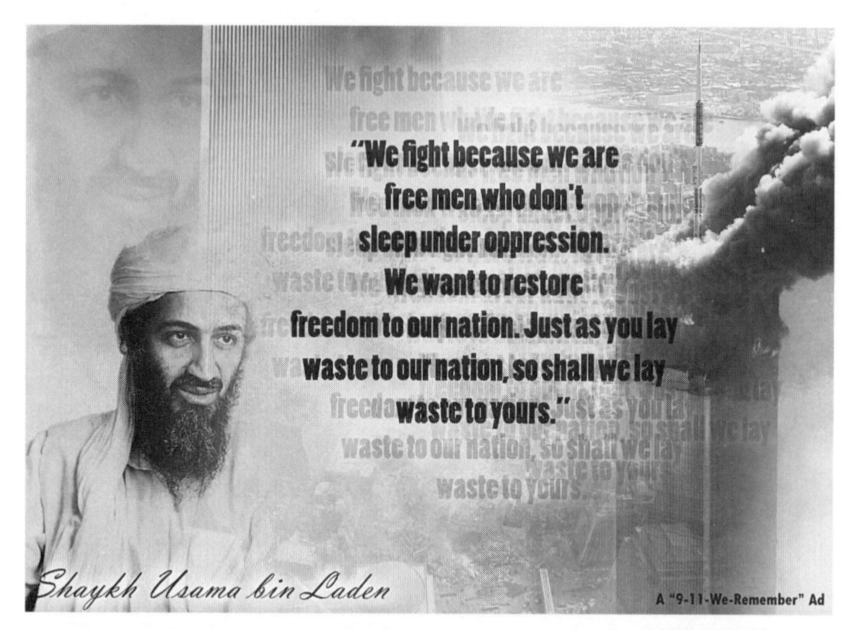

"We fight because we are
free men who don't
sleep under oppression.
We want to restore
freedom to our nation. Just as you lay
waste to our nation, so shall we lay
waste to yours."
Shaykh Usama bin Laden
A "9-11-We-Remember" Ad

Every year, Taliban texts in various languages commemorate the 9/11 attacks. In this image, a picture and quotation of Osama bin Laden is featured alongside a burning World Trade Center. The term "nation" here refers to a community of believers—Muslims—not any particular nation-state. The "nation" mentioned by bin Laden is that of the "Crusaders," a disparaging term referring to Christians. Taliban authors assume that Americans are Christians without considering the multicultural nature of the United States.

Taliban authors portray the United States as a contemporary Pharaoh to describe American oppression and justify militant jihad. Taliban authors see themselves as inheritors of a legacy in which Moses defeated the Egyptian Pharaoh and Jesus defeated the Roman Pharaoh in a long line of Semitic prophets standing against political oppression through political and moral solidarity. Among these images, specific American symbols exemplify general Pharaoh-like tendencies: the American army exemplifies brute force, the image of CNN exemplifies disinformation, a baseball stadium exemplifies lack of political consciousness, Dick Cheney exemplifies falsehood about Iraq's nuclear capabilities, Malcolm X exemplifies political assassination, and President Obama with the Saudi King exemplifies bribery and corruption within Islam.

look for) the Ilah (God) of Musa; and verily, I think that he [Musa] is one of the liars." [28:38]

"And (with) Pharaoh, who had pegs (who used to torture men by binding them to pegs)?" [89:10]

For this reason, he said:

"... O chiefs! I know not that you have an ilah (a god) other than me..." [28:38]

"..."I am your lord, most high", [79:24]

There is no doubt that pharaoh was a down-trodden and spoiled individual... He used to worship statues and cows, and a pearl that he used to have hung. After all this, what else can his claims of lordship and godhood imply?

Some of the people of knowledge say that Pharaoh was an atheist!

Maybe this means that he became so arrogant in his sinfulness that he eventually did become an atheist - because nothing occurred contrary to his will during his life. Nor did he remain ill for any significant period of time – until the caller of truth - Musa ﷺ - came to him with his weaponry:

"... (These are) among the nine signs (you will take) to Pharaoh and his people, they are a people who are the Fasiqun (rebellious, disobedient to Allah)." [27:12]

As one poet said:

He was a man
From the army of satan
Until he advanced so much
That Iblees became his disciple!

Yes, some matters did go against his will, like the upbringing and freedom of Musa ﷺ:

"Then the household of Pharaoh picked him up, that he might become for them an enemy and a (cause of) grief. Verily! Pharaoh, Haman and their hosts were sinners" [28:8]

Similarly, there was the matter of Musa's ﷺ freedom from him when Musa ﷺ called him towards *Tawheed Al-Hakimiyyah* (Monotheism in Legislation), but Pharaoh was a ruthless individual. As Allah The Exalted said:

"The chiefs of Fir'aun's (Pharaoh) people said: "Will you leave Musa and his people to spread mischief in the land, and to abandon you and your gods?" He said: "We will kill their sons, and let live their women, and we have indeed irresistible power over them." [7:127]

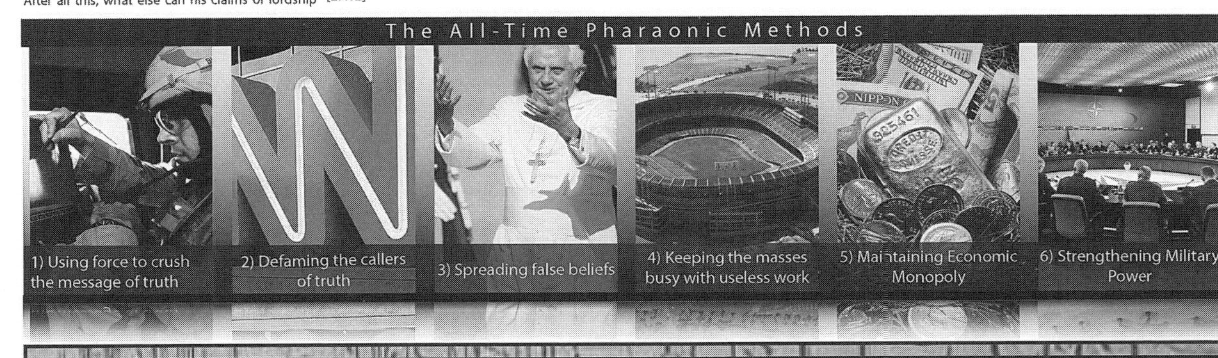
The All-Time Pharaonic Methods
1) Using force to crush the message of truth
2) Defaming the callers of truth
3) Spreading false beliefs
4) Keeping the masses busy with useless work
5) Maintaining Economic Monopoly
6) Strengthening Military Power

Some Words Regarding the Means and Ways by which Pharaoh Protected his Government

Mufti Abdullah Muhammadi (HA) writes:

The Pharaohs of every day and age use similar measures to protect and strengthen their Government. Let us take a brief look at them:

The Quran explains in detail that Pharaoh utilized oppression, force, persecution, reassurance and greed to strengthen the hold of his Government. He enslaved one group of people, bought another (with bribes) and kept another in ignorance and misguidance (by lying and falsehood). These measures are undertaken by almost every tyrant:

Allah The Exalted said: "Have they (the people of the past) transmitted this saying to these (Quraish pagans)? Nay, they are themselves a people transgressing beyond bounds (in disbelief)!" [51:53]

First Method of Subjugation: Killing, Imprisonment and Humiliation

Pharaoh's oppression of his people is not hidden to anyone. Tyrants of every age forcibly enforce their laws and systems upon the people in order to enslave them. If some of the people of truth reject servitude to him, then their last measure is always the same: Oppression... But they seldom realize that victory belongs to the people of truth:

Allah says: "And indeed We saved the Children of Israel from the humiliating torment, From Pharaoh; Verily! He was arrogant and was of the Musrifun (those who transgress beyond bound in spending and other things and commit great sins)." [44:30-31]

"And (remember) when We separated the sea for

you and saved you and drowned Fir'aun's (Pharaoh) people while you were looking (at them, when the sea-water covered them)." [2:50]

Second Method of Subjugation: Keeping the masses ignorant and misguided

Ignorance is the antithesis of knowledge. Pharaoh ordered the people to obey him in his arrogance. This is because an arrogant and insolent person wants that other people should also become like him, as Allah says:

"They wish that you reject Faith, as they have rejected (Faith), and thus that you all become equal (like one another)…" [4:89]

The Arab people consider ignorance as misguidance and vice versa. [Tabari: 19/67]

Ignorance is lack of knowledge. It also refers to holding a belief about something that is contrary to its reality. One of the signs of the Day of Judgment is the decline of knowledge and the rise of ignorance as the Hadith of Prophet ﷺ states in Bukhari:

"Indeed, from the signs of the Hour is that knowledge will be raised, ignorance will appear (and become widespread), fornication will spread, alcohol will be imbibed, men will depart, and women will remain, to the extent that for 50 women, there will be a single guardian." [Bukhari, Muslim]

Sayyid Qutb Shaheed (RA) writes that the Pharaohs of every age block the ways which would make people recognize the truth. This occurs until people forget the actual knowledge and then cease to return to it or struggle for it. The Pharaohs and their followers of every age misguide people as they please. This is so that they can enslave them easily. It is easy to rule over a people who lack true knowledge. (As can be witnessed today)[4]

"Thus he [Pharaoh] befooled and misled his people, and they obeyed him. Verily, they were ever a people who were Fasiqun (rebellious, disobedient to Allah)." [43:54]

The exegetes write that he had kept his people ignorant and that they obeyed him due to their

6) Lying and Falsehood
7) Political Assassinations
8) Oppressive Taxing
8) Bribery
9) Considering people as slaves
10) Answering Logical proofs by force

ignorance and foolishness.[5] The Tawagheet and the evil scholars of today are doing exactly this.

The All-Time Pharaonic Methods

The Pharaoh used the following measures to deprive the people of their rights:

1) Crushing the message of truth by force

Pharaoh prevented the people from freedom of belief and did his utmost to crush the message of truth via imprisonment and killing so that the light of truth would not reach the masses. Various evidences from the Quran testify to this:

"And We inspired Musa and his brother (saying): "Take dwellings for your people in Egypt, and make your dwellings as places for your worship, and perform As-Salat (Iqamat-as-Salat), and give glad tidings to the believers." [10:87]

Allama Nasfi Hanafi (RA) says that when the Children of Israel experienced danger due to following the truth, they were ordered by Allah to pray secretly in their houses so that the Kuffar would not put them to trial because of their religion, just as the Prophet ﷺ and his Companions ؓ did (pray secretly) in the beginning in Makkah. [Tafseer Nasfi 2/139]

Are not the Pharaohs of today doing exactly the same thing?

2) Defaming the callers of truth

Pharaoh defamed the caller of truth Musa ؑ so that the people would not get close to him. He labelled Musa ؑ as a magician and told the people that he (Musa ؑ) wanted the kingdom for himself.

… And as the Kuffar of Makkah said to the Prophet ﷺ: "And those who disbelieve say: "Listen not to this Qur'an, and make noise in the midst of its (recitation) that you may overcome." [41:26]

And the Pharaohs and Abu Jahls of today are doing the same thing with the Mujahideen today so that the people do not understand their call of establishing the Rule of Allah in the world and cutting the roots of this secular, democratic system. This is because the hypocrites and the apostates know that if the masses come to know about the message of the Mujahideen, then truth and falsehood would be laid bare. There is little need to go into detail.

3) Spreading false beliefs & keeping the masses busy with useless work

As Allah The Exalted says: "And so the sorcerers came to Pharaoh. They said: "Indeed there will be a (good) reward for us if we are the victors." He said: "Yes, and moreover you will (in that case) be of the nearest (to me)." [7:113-114]

Just as Pharaoh utilized the sorcerers to keep supreme his authority (and they complied with him in order to gain worldly benefits), tyrants of every day and age keep such people and armies

4. [In the Shade of Quran: 5/340]

5. [Qurtubi, Fath-ul-Qadeer, Rooh-ul-Mu'aani, Baidhawi]

CHAPTER 5

International Relations in the Virtual Emirate

THE TALIBAN'S REPRESENTATIONS of the in-group as all Muslims and out-groups as infidels raise questions as to whether these categories orient its international outlook. One clue may be the Taliban response to foreign events. In his 2013 State of the Union address, President Obama announced the end of combat operations in Afghanistan: "This spring, our forces will move into a support role, while Afghan security forces take the lead. Tonight, I can announce that over the next year, another thirty-four thousand American troops will come home from Afghanistan. This drawdown will continue. And by the end of next year, our war in Afghanistan will be over" (Obama 2013). The Taliban immediately criticized the continued presence of American troops: "The Americans, who are renowned in the world as the blood suckers and professional murderers of the oppressed masses, want to keep the war-flames alive in the war-ravaged Afghanistan, so that they could take the revenge through this war from the Afghans who have inflicted devastating corporeal and financial losses on them in the battle field" (Super User 2013).

This post indicates that the Taliban views the United States as an enemy through such stereotypes as "blood suckers" and "professional murderers." We know from chapter 2 that Mullah Omar promoted a conception of non-Muslims as constituting out-groups. We also know from chapters 3 and 4 that those in an in-group may stereotype out-groups based on social competition. To what extent is there a relationship between Taliban in-group/out-group representations and attitudes toward nation-states? How can cultural psychology and psychiatry extend our understanding of international relations? This chapter investigates Taliban constructions of nation-states to explore psychological mechanisms for depicting insiders and outsiders. I first review theories on culture and cognition, focusing on image theory in psychology. I next analyze Taliban texts for images of nation-states. This work can help us investigate the Taliban's self-positioning as a voice for global Islamist movements in the virtual emirate, with implications for foreign policy.

CULTURE AND COGNITION IN INTERNATIONAL RELATIONS

The challenges of constructing a unitary understanding of abstract nation-states, with their thousands of institutions and countless inhabitants, have generated distinct theories and methods incorporating cognitive psychology within international relations. Just as humans simplify decision making in everyday life to the most relevant tasks, without deliberating over all sensory input from the environment, foreign policy experts draw interpretations from the actions and statements of relevant foreign leaders in a world replete with danger and uncertainty (McGraw 2000; Berejikian 2002; Rosati 2000; McGraw and Dolan 2007). Such simplification occurs through shortcuts in cognitive information processing, which lead to interpretations based on views and attitudes that are steeped in culture and history (Renshon and Renshon 2008). For example, President Obama's speech elicited a cognitive shortcut about all Americans from a Taliban author who made a broad generalization rooted in the recent history of the American–Afghan conflict. Culture molds all steps in information processing, since shared understandings and expectations are used to communicate interpretations of others' actions (Adler 1997; Ross 1995; Ross 1997; Houghton 1998; Glenn et al. 1970). The Taliban author assumes that his readership would share an understanding of Americans as world-renowned for military aggression and political oppression, leading to a negative interpretation of the United States.

If cognitive psychology theorists have highlighted the role of culture in drawing interpretations about nation-states, social psychology theorists have applied social identity theory to international relations. One type of interpretation is the image, first advanced by Kenneth Boulding (1959), who argued that "images which are important in international systems are those which a nation has of itself and of those other bodies in the system which constitute its international environment" (120–121). Political elites construct images to simplify the environment, evaluate information, and build a worldview (M. Cottam 1985), classifying other nation-states as friendly (in-group) or hostile (out-group) along relevant dimensions based on national interests (Hermann, Isernia, and Segatti 2009). Images can be studied through texts, such as speeches, historical archives, and policy documents (R. Hermann 1985). Analyses of multiple images for the same nation-state can uncover intragroup variations, image prototypes that triumph over others, and the power of elites responsible for such prototypes (Alexander, Levin, and Henry 2005). Conceding that the Islamic Emirate of Afghanistan is not a nation-state and that images may differ by language, I believe that analyses of images from individual authors in the virtual emirate can uncover which image prototypes prevail in Taliban group discourse.

The scholarship of Richard Hermann and Michael Fischerkeller is noteworthy for linking images of nation-states to a range of strategic foreign policies. Hermann and Fischerkeller (1995) define the image as "a subject's cognitive construction or mental representation of another actor in the political world" (415). Because no model can account for the many factors that affect decision making or predict how political elites view the world, studying images may illustrate how elites represent situations and make choices (Hermann and Fischerkeller 1995). Along with other image theorists (R. Cottam 1977; M. Cottam 1992), Hermann and Fischerkeller (1995) propose five unique images based on (1) perceptions of mutual threat or opportunity, (2) relative military power, and (3) judgments about the cultural sophistication of decision-making behaviors among political elites:

Enemy: An image of a state as highly competitive, comparable in power and cultural sophistication. Both actors assess threats similarly and wish to eliminate each other. The other is viewed as evil and untrustworthy, leading to policies of deterrence and containment.

Ally: An image of a state as possessing similar goals and interests, comparable in power and cultural sophistication. The other is viewed as benign and motivated by altruism as much as self-interest. Both actors cooperate for mutual gain.

Degenerate: An image of a state as stronger in power but inferior in cultural sophistication. Leaders seek to preserve present gains rather than articulate a vision for the future, eliciting insecurities about unpredictable actions and leading to policies of avoidance.

Imperialist: An image of a state as superior in power and cultural sophistication. The other is perceived as pursuing economic exploitation by importing raw materials or demanding a market for goods and services, leading to rebellion.

Colony: An image of a state actor as inferior in power and cultural sophistication such that the more powerful country can exploit the weaker one. This image leads to exploitation of the other to increase one's gains.

Hermann and Fischerkeller (1995) demonstrate the validity of these images by analyzing every statement of two or more pages from the three most powerful leaders of a nation-state in several case examples. This methodology assumes that the international system rests on the nation-state as a unit of analysis and that each nation-state possesses multiple leaders, though I have shown that neither condition applies to the Taliban as a non-state militant movement headed by a single supreme leader. A focus on nation-states may ignore non-state actors—such as the Taliban and other militant groups (Al Qaeda and the Islamic State, among others)—that have indelibly influenced the international system. Since Taliban leaders do not regularly issue statements, I analyze only those texts from Taliban authors that address specific policies of nation-states. I included countries in this analysis if three or more articles by Taliban authors addressed specific country policies rather than mentioning them in passing; to take the first country, for example, I included Burma because Taliban authors have written entire articles about Burmese government policies. In contrast, I have excluded a country like Malaysia, which has been listed only in passing within articles addressed to the entire Muslim world. Based on chapter 3's analysis of in-groups and out-groups, I hypothesize that only nation-states with large Muslim populations will be viewed as allies.

Burma

Taliban authors depict Burma through an enemy image for atrocities against Muslims. A statement in *Sharīʿat* from the Islamic Emirate of Afghanistan explains the threat: "Men, women, and children are being roasted in fire like animals without mercy which is not permitted in any law.... They are being removed from their settlement areas and homes, their properties are being usurped, and their honor and dignity are being violated" ("Barmī Musalmānon Kē Khūnī Sānhe Kē Bāre Men Imārat-e Islāmīya Kā Elamīya" 2012). Hassan Momand (2012) details Burma's total domination of Muslims through such phrases as, "Without the government's permission, Muslims do not have the right to move from one place to another" and "Muslims do not have the right to have more than three children" (23). Both examples level charges of government interference in the everyday lives of Muslims.

Taliban authors perceive these evils as a result of a non-Islamic system of government. Mustafa Hāmid (2012b) demonstrates a cognitive shortcut in forming in-groups and out-groups along the dimension of religion, generalizing the Burmese perpetrators as Buddhists: "The discourse of decadent Western liberalism that the world raised upon the destruction of the Buddhist statues [in Afghanistan] and silenced upon the Buddhist killing of thousands of Muslims is because stone is more precious to them than the blood of Muslims" (7). Hāmid (2012b) ridicules international law for its subservience to American interests, linking events in Burma and Afghanistan: "Do the Rohingya people, the Afghan people, and the Taliban movement lack recourse and a law based in reality? Are this recourse and this rare law of reality not provided except through the embrace of the American colonizer?" (7). Inamullah Habibi Samangani (2014) criticizes democratic reformists for not halting the slaughter of Muslims: "For democrats, only democracy is dear. It is possible that the democrats of the world are not aware that today, the Rohingya Muslims of Burma are definitely crawling to "minority" [status] to the extent that they are giving up their lives in tents" (26). The only solution is Muslim solidarity: "The voice of the Islamic Emirate that reaches the world's corners and awakens populations, governments, organizations, and individuals is only the result of divine assistance, and we can say it is a blessing from the sincere efforts of the Islamic Emirate for the weak Muslims of the world" (Hikmat 2012a, 10). These authors position the Taliban not just as a militia implementing *Shari'a* law in Afghanistan but also as a global champion of Muslims everywhere.

Canada

Taliban authors represent Canada as an imperialist. Samiullah Zurmati (2014) specifies Canada's colonizing ambitions: "The Western occupiers divided all of the provinces of Afghanistan among themselves: Kandahar to

Canada, Helmand to Britain, Ghazni to Poland, Uruzgan to Australia and Holland, and the responsibility for the remaining provinces was entrusted to American forces. This division of the occupiers reminded the Afghan people of colonial times" (27). Zurmati invokes nationalism and shared understandings of Afghanistan's colonial past as a cognitive shortcut to explain Canada's present actions.

An imperialist image leads to armed resistance against the foreign power and native collaborators (Hermann and Fischerkeller 1995). Qari Habib (2014) uses the Taliban's armed resistance to interrogate Canadian foreign policy:

> What did Canada accomplish from the war in Afghanistan? The Canadian officials announced on their day of retreat from Afghanistan that its losses in Afghanistan were 158 casualties and thousands injured, in addition to the killing of Canadian advisors, high-level diplomats, and a journalist. However, neutral statistics indicate that the losses of Canadian forces were many times more than what Canadian officials announced. Observers of military positions in Afghanistan affirm that Canadian forces did not accomplish anything in this country that they can present to the Canadian people. (5)

Habib's passage displays a recurrent cultural theme in Taliban discourse of distrusting the narrative and statistical accounts of foreigners. He also contrasts Canada's superior military power and decision-making sophistication with the will of its people. Habib (2014) concludes with the impact of Canada's retreat on Afghan collaborators: "Many of the Afghan translators and contractors in Canadian positions and offices within Afghanistan carry Canadian citizenship. The Canadians came with them to Afghanistan with assurances of safety. These are individuals in high-level government like the current governor of Kandahar and others like him who reached these positions with the support of the Canadians. . . . And they demand the carrying out of promises and assurances that were presented to them" (5). The imperialist image of Canada permits authors to present the Taliban as patriots defending all of Afghanistan rather than as a movement solely concerned with Sunni or Pashtun interests.

Egypt

Taliban authors depict the people of Egypt as an ally but the government as a colony of foreign powers. "Misr Ko Shām Jaisē Hālāt Men Dakhalnē Kī

Sāzishen Aur Misrī Āwām Kī Zimmedārīyān" (2013) praises the 2012 election of the Muslim Brotherhood's Mohamed Morsi (b. 1951) after the overthrow of Hosni Mubarak (b. 1928):

> In the whole Islamic world, and especially the Arab world, there was a demonstration of joy that after the end of military rule for 61 years and personal rule in Egypt, for the first time the people found an opportunity to participate in power. Election results made clear that Egyptian Islamic parties are not just pressure groups on the armed military but rather friends of the people and parties that have a special love for the people. (2)

This unnamed Taliban author uses Morsi's election as a way of relating to Egyptian Muslims along the relevant dimension of religion.

Morsi's downfall provides an occasion to comment on Middle Eastern policy. The same author writes of a foreign conspiracy: "From historical and geographical perspectives, neither America, Israel, or other Western governments, nor did the supporters of Hosni Mubarak in Egypt and other secular parties want the success of Islamists in the extremely important country of Egypt" (2). The author views Morsi's overthrow through the lens of religion: "We understand President Morsi is being given a punishment for Islamism, Muslim unity, sympathy for Palestinians blockaded in Gaza, steps to establish a balance of power in the Middle East against Israel, and taking steps toward self-determination in removing Egypt as America's instrument of power" ("Misr Ko Shām Jaisē Hālāt Men Dakhalnē Kī Sāzishen Aur Misrī Āwām Kī Zimmedārīyān," 2). Author Walid Afghan (2013a) identifies domestic opponents to Morsi as "Coptic Christians in Egypt and domestic and international, secular, liberal, and independent thinking elements," explaining that "on July 4, the military announced the removal of Egypt's elected president Dr. Muhammad Morsi from office, and in his place, named the country's chief justice Adli Mansur as the country's interim president who is a Coptic Christian" (14). Both authors consolidate Egyptian identity around Islam, and Afghan displays a cognitive shortcut by stereotyping domestic and foreign opposition to Morsi through Christianity. For this reason, Shahid Zui (2013b) comprehends Morsi's overthrow as an international call for jihad: "We are certain that in the current event, through the blessing of the great sacrifices of Muslims, the world's and Egypt's Muslims are coming much closer to their religion. The passion for sacrifice and jihadist thought will be stronger upon coming closer to religion" (16). Taliban authors exploit Morsi's brief stint in power to position themselves as the vanguard of the Muslim world for rallying support against non-Muslim domestic and foreign powers in the Middle East.

England

Taliban authors depict the British government as imperialist, intent on subjugating Muslims at home and abroad:

> It can be expected that the British government will carry on with its current policies. It will not abandon its neocolonial foreign agenda in the Muslim world. This will result in continuing anger amongst the Muslim population in Britain, some of whom will take heed of the Mujahideen's call to attack the Crusader nations in their own lands. This increasing threat will be countered by more draconian legislation by the British government; the media will continue to demonise Islam and the Muslims and far-right organization[s] like the EDL [English Defence League] will be allowed to attack Muslim areas. (Muhajir 2014, 17)

The use of terms with the root *colonial* recurs in an official statement commemorating the ninety-third anniversary of Afghanistan's freedom from British rule: "The English colonizer, who had controlled many countries of the world and the region, accepted the freedom and self-determination of Afghanistan. We present a blessed offering to the Afghan mujahid nation on the arrival of this historic day. May this freedom obtained from the English colonizer be blessed, and may God also ordain that freedom be obtained from the current American manslaughter" ("Angrez Istimār Se Mulkī Āzādī Kī 93vīn Sāl Mukammal Honē Par Imārat-e Islāmīya Kā Elamīya" 2012). An article by Saiful Adil Ahrar (2012c) contends that the British joined forces with the United States to reestablish imperialist glory: "At that time, the British colonizer suffered the worst loss in Afghanistan's Helmand province. For this reason, once the United States attacked Afghanistan, then Britain built a large military base after having sent its military to Helmand under the pretext of taking revenge for its forefathers" (19). Each Taliban author exploits the legacy of colonialism as a cognitive shortcut to explain British interference in Afghan affairs and its cooperation with the United States.

France

Taliban authors depict France as an imperialist power. Abu Ahmad (2011) describes an account from a villager detained by French and Afghan forces: "The French tyrannized us a great deal. Therefore, the people of the province hate them, although what they undertook by way of violence and subjugation toward the innocent villagers was still better—despite their injustice and assault—than the enslaved Afghan forces that aid the occupiers, torture prisoners, and take a large amount of American dollars" (13). Dari author Zaheer Khan (2012) also describes French policy in Afghanistan: "A few days ago in the province of Kapisa in Togab district among the friends of jihad and the mujahideen, one Afghan mujahid killed ten French soldiers with bravery, for which reason the head of the French president Sarkozy was so low from dishonor that he immediately announced the withdrawal of French troops from Afghanistan before the determined deadline (41)." In the Pashto language, *sarkozy* is used to refer to a person whose head is low. Here, culture supplies a linguistic pun to equate the French president's name with shame through a cognitive shortcut.

Taliban authors also decry French policy toward its Muslim citizens. Saiful Adil Ahrar (2012b) decries Sarkozy for outlawing headscarves:

> During the rule of Sarkozy's government, a ban was instituted upon Muslim girls wearing scarves and Muslim women wearing veils in schools and colleges, whose example is not found in any law of the world. After the passage of the bill against Islam, the Sarkozy government imagined that Muslim women living in France would stop wearing veils from fear of punishment. However, the government had to confront an intense disappointment when some Muslim women came out in public wearing veils without worrying at all about punishment. (32)

Another article in *Srak* parodies France's reputation for human rights:

> The pattern of human rights in France—a country that European history recalls with the title "The Cradle of Freedom and Democracy"—has shown its true colors through evidence such as the influence of Zionist lobbies in government, the insistence of the state on keeping its secular nature, and state capitalist domination over all. Religious minorities such as Muslims and ethnicities such as those with Arab ancestry and Blacks not only are

> deprived of parliamentary seats and other decision-making institutions, but are also taken as targets by racists and extremists. ("Qānūn-e Mana' Hejāb Dar Gharb wa Ta'āruz-e Ashkār-e An Bā Usūl-e Demūkrāsī" 2011)

Taliban authors portray the French government as an ideological enemy subjugating Muslims in Afghanistan and France. Shared understandings based on colonialism ("enslaved Afghan forces"), language, and religion enable authors to position the Taliban as a protector of Muslim rights against France whether in Afghanistan or beyond.

India

Taliban authors envision Indian Muslims as allies and the government as an enemy. Maulana Asim Umar (2013b) laments the absence of Indian Muslims among the mujahideen: "At such a time, the global Jihad leadership feels justified in asking the Indian Muslim scholars and masses as to why the Jihad battlefields remain deprived of their blessed presence" (42). Umar (2013b) reminds them of past glories:

> O people who ruled India for 800 years! O you who uphold the light of Tawheed [unity] in the midst of the darkness of Shirk [polytheism]! If the Muslim of the world has awakened today, then how can you remain sleeping? If the Muslim youth of today has entered the battlefield with the slogan of "Shariah or martyrdom!", then how can you remain behind? Your fathers fought till the last to save the Khilafah Usmania [Turkish Caliphate]. (45)

Umar (2013b) establishes a marital lineage for Indian Muslims: "Oh inheritors of Muhammad bin Qasim and Mahmood Ghaznavi! Children of Aurangzeb and Abdali![1] Stand up! Stand up to establish the Khilafah in the world!" (46). Umar (2013b) discounts that Aurangzeb and Abdali ruled competing empires; they belong to the same in-group by virtue of their religion and symbolize Muslim military might in South Asia. Consequently, Walid Afghan (2013b) suggests that rapes publicized in 2013 reflected India's turn away from Islam:

> History is witness that when there used to be the rule of just Islamic law upon the territory of Hindustan, then no such incident ever occurred year after year. Even today, if the Indian people and government could adopt

> Islamic laws and just administration, then surely again, those scenes of gardens and springs of peace and tranquility in India would become visible. If this is not possible for them, then at the least, if the Indian government and people come out of religious fanaticism and counter the causes of immorality, there can be a significant reduction in these events. (7)

Afghan (2013b) terms India *Hindustan*, literally "the place of the Hindus," to emphasize its Hindu majority population and the moral decay of a non-Islamic polity. Abid Tanwir (2014) chastises the Indian government for supporting a security agreement allowing American troops to remain in Afghanistan after NATO forces withdraw:

> India is the one country that has never remained neutral in Afghanistan's problems and internal affairs. It has always tried to end Afghanistan's religious and national power and cooperated with the killers of Afghans. After the Russian aggression of Afghanistan, the whole world supported the subdued and oppressed Afghans. India was the country that not only supported the violations of the Red Army, but also maintained strong political and military support for the Communist administration in Kabul to the final day. (6)

Tanwir constructs India as an infidel out-group to strengthen an Afghan identity based on religion.

Israel and Palestine

Taliban authors depict Israel through an enemy image as powerful and threatening to Arab Muslims. Shahid Zui (2012) contends that theology motivates Israeli politics:

> According to Jewish beliefs, God (may He be exalted), even now, converses with his chosen, and those chosen (rabbis) are the religious leaders of the Jews. Their teaching is that God gives them orders to go forth and kill Arabs. The religious leaders of the Jews issue fatwas related to this exact belief in which the manslaughter of Palestinians is not only permitted, but it is also said that God (may He be exalted) is happy from this action. (34)

Shared perceptions about Jews in Taliban discourse permit Zui to stereotype Israelis as an out-group based on religion.

In contrast, Taliban authors portray Palestinians as allies, subject to oppression from Western powers. Badruddin Hammadi (2007) writes: "History will inform us that the Hamas movement is uniquely skilled and unprecedented in benefitting from the mistakes of past revolutions. The Palestinian revolution today is in safe hands, praise God, and it would not be strange that history would later register thanks for this jihad movement which returned its philosophy and splendor to the revolution" (30). Hammadi compares Hamas to the medieval warrior Salahuddin (see chapter 3), who "would not have been able to triumph over the Crusaders in Palestine and expel them from it if he had not first undertaken war with the Arab emirates which had been subservient to the Crusaders and supported them" (30). Hammadi draws a historical analogy, connecting past with present, to advance a martial identity based on religious unity against non-Muslims. This unity recurs in an official Taliban statement in *Al-Somood*:

> Oh mujahideen brothers in Palestine! We consider the Palestine issue and the Holy Al-Aqsa issue to be our issues. And we consider their defense and protection to be among our religious obligations and Islamic responsibilities. . . . Even if our hands do not reach your shedders of blood among the Zionist murderers, we are here in Afghanistan and in our trenches of war fighting the Americans who are the real defenders of Israel. We take revenge for your martyrs and we undertake and execute military attacks against them. ("Al-Munāsara Al-Muqāwama Al-Falistīnīya Farīdha Dīnīya Wa Masūlīya Islāmīya" 2008)

The Taliban embraces the Palestinian issue to such an extent that it claims revenge for slain Palestinians by killing Americans in Afghanistan. Terms such as "we" and "our" forge kinship bonds through religion. Mahmood Ahmad Nawid (2014) addresses Palestine:

> Your news is the news of sorrow of an oppressed nation, your news is the news of a wound old and deep, your news is a reminder of the difficult days, your news is the news of Muslims who pray daily for wanting respect and demanding freedom only for the crime of Islam and religion, your children are being killed only for this reason, being burned, their red blood is being shed, their houses and huts are being deserted and destroyed. (9–10)

Nawid attributes Palestinian oppression to the inability to practice Islam with dignity, shifting the Israeli–Palestinian issue from a geopolitical conflict to a religious war, akin to the Taliban's perceptions of Afghanistan.

Mali

Taliban authors portray the people of Mali as allies against a colonial government beholden to Western powers:

> Mali is part of the area that the caravans of the early Muslim generations freed from Jahiliyyah (ignorance). It is a country spanning an area of around 1.25 million square kilometres and having a population of up to 14.6 million out of which more than 90% are Muslims. It is blessed by Allah with a great many natural resources such as gold, uranium among others. In fact, it possesses the 8th largest gold reserves of the world. But sadly, it has been subjected to the same old secular onslaught of the Kuffar [infidels]. The Government is secular and has carried out the plans of the [W]estern masters quite ably. This relatively unknown country caught the eye of the world when the lions of Islam established Shariah in its key areas; this proved as a thorn in the throats of the European tyrants whose thrones shook at the establishment of the Shariah so close to their waters. (Salahuddin 2013, 57)

Hassan Momand (2013a) writes: "The event of September 11th happened and the whole Western world along with the United States found an excuse to come close to Central Africa" (9). Momand (2013a) explains France's long-standing colonial ambitions: "In 1880, French forces were able to enter Mali and made this country a colony that was maintained until 1940. This colonization lasted a long time. During this period, the language, culture, and religion of the poor people of Mali should have all changed, but let there be praise that Mali's Muslims did not change their firm belief despite the intense efforts of Christian missionaries and the extreme pressure of the French government" (10). Salahuddin and Momand differentiate the Muslim population from a series of treacherous governments to establish in-groups and out-groups by religion. Taliban authors allege economic and cultural exploitation, conflating Christian interests with European colonial interests that can be dispelled by a firm sense of Muslim identity.

Pakistan

Pakistan elicits multiple images. The theme of Pakistan as a nation-state founded on the basis of Islam but now attacking Muslims supports a degenerate image:

> They [nationalistic armies] have proven to be in reality, protectors of this irreligious "new world order" that is imposed upon the Muslim Ummah. A case in point is the Pakistan Army which proudly claims to be the "defender of a Muslim nation" and yet recently proclaimed on International Media that, "We have more threat from the Western Afghan border (Mujahideen) than from (Mushrik, polytheist) India." In reality, this statement translates to nothing but, "The Mujahideen who want to worship Allah Alone and perform the noble act of Jihad in the Path of Allah are more of a threat to us than these Mushrik Hindus." What a sad state of affairs! What a far cry from what Islam stands for! Is this a Muslim army that seeks the establishment of Islam or is it standing in the way of those who wish to perform this sacred duty? (Anwar 2013b, 30)

Adnan Rasheed (2014b) continues this theme of an apostate Pakistani government by condemning Pakistani military action in North Waziristan:

> Cobra gunships and machine gun-fitted Mi-17 helicopters targeted the homes of innocent civilians killing more than 70 men, women and children and severely injuring more than 100. Anyone who breathed was declared an enemy combatant. The Pakistan Army continued the operation for a week; homes, shops, mosques, markets, schools and hospitals all were targeted. The Army brutally and ruthlessly bombed down the main Mir Ali *Jami'a Masjid* located in the heart of the city, killing 6 worshippers in the process. (26)

Anwar and Rasheed contrast the might of the Pakistani Army with its attacks on civilians. Both authors also emphasize an in-group through religion, with Anwar equating a blasphemous Hindu identity with archrival India, and Rasheed decrying the bombings of civilian institutions, such as mosques.

In contrast, Pakistani religious scholars are portrayed as allies. An unattributed article in *Srak* praises their abstention from a religious conference that was seen as a Western conspiracy to support the official Afghan government (see chapter 4):

> The coming conference planned between scholars of Afghanistan and Pakistan is nothing but an arrogant international conspiracy, and they do not have any great hopes for it. They have imposed two conditions for participating in this conference; that the armed mujahideen of the trench (representatives of the Islamic Emirate of Afghanistan) participate in this

> conference to explain their position to everyone. And the second condition of these scholars has been to restore the good name of the mujahideen sitting in the trenches since the colonizing crusaders and their regional slaves want this group of socially righteous people to be defamed by various names. ("Konfarans-e Ulemā Qabl Az Iniqād Nākām Gardīd" 2013)

Religious scholars are represented as bold defenders of the mujahideen against an international conspiracy. Similarly, Adnan Rasheed (2014a) regards the people of Pakistan as allies: "[The] Mujahideen are the only defenders and sincere friends of the people of Afghanistan, Pakistan, India, Bangladesh, and others. The Muslims of the subcontinent and Afghanistan must help and join their true friends if they want to save their Deen [religion], honor and lives" (23). Taliban authors position the mujahideen as a righteous force on the side of civilians and religious scholars against a savage Pakistani government that has strayed from the path of Islam.

Russia

Taliban authors depict Russia as an imperialist force. The Taliban has officially released an annual statement to commemorate the defeat of the Soviet Union. In 2008, the statement in *Al-Somood* praised "the Afghan Muslim people, who supported the freedom and defense of proud Islamic laws against surrender to colonial forces or bowing their heads to inimical tyrants, raised the banner of jihad against them, and revolted against these invading forces which were almighty and Pharaonic for its time" ("Bayān Imāra Afghānistān Al-Islāmīya Bi-Munāsaba Yaum Ihtilāl Afghānistān Min Qabla Al-Quwwāt Al-Rūsīya" 2008). Religion provides a cognitive shortcut to form an in-group based on Islam against an infidel out-group. The symbol of the Soviet Union as a contemporary Pharaoh, flush with modern weaponry, recurs in an official statement the following year: "The tyrannical Soviet forces that were heavily armed with the latest types of developed weapons were forced to flee from Afghanistan at a time when no success was possible during its ten-year occupation. In exchange for success, more than 15,000 of its soldiers were killed and thousands of military equipment were destroyed just as the financial value of billions of dollars was ruined" ("Bayān Imāra Afghānistān Al-Islāmīya Bi-Munāsaba Dhikra Al-'Asharūn Li-Insihāb Al-Quwwāt Al-Sofitīya Min Afghānistān" 2009).

Apart from commemorating the Soviet Union's failed Afghan campaign in the 1980s, Taliban authors warn against Russia's contemporary imperialism. Janbaz (2013a) sees Russia, the United States, and NATO as imperialist forces: "To prevent any Islamic states from reviving, Russia, America, and NATO have taken turns in playing their satanic (*shaitānī*) roles. When one tires, then another takes its place. First Russia, then NATO, then America. . . . And now Russia again is weighing interference" (29). Janbaz (2013a) attributes Russian imperialism to the fear of Islamism in Central Asia: "Where Stalin's statue used to be saluted, mujahideen forces there—having received help from the Taliban—can wave the flag of jihad and freedom. In this way, it is very possible that Russian and American forces would collapse in the upper states. And that the days of freedom and nights of liberty become available to the Caucasus" (29). History provides a shared understanding to emphasize Russia's long-standing sinister motives as Taliban authors proudly announce support for mujahideen outfits in Central Asia.

Syria

Similar to Mali and Pakistan, Taliban writers assign opposite images to the people and government of Syria. Maulana Asim Umar (2013a) depicts the people of Syria as an ally and the government of President Bashar al-Assad (b. 1965) as an enemy:

> The sisters and daughters of the Sunnis have been taken away and subjected to horrible abuses in the jails. These Shia who are probably the worst of creations in the entire universe, the killers of Umar, Usman, Ali, Hassan and Hussain,[2] have targeted the pious daughters of the Ahlus-Sunnah [the Sunni people] out of their historical spite. Maybe, with it, they intend to cool the hatred that their mentor ibn Saba[3] had planted in their hearts a long time ago. Or maybe they intend to seek nourishment of their faith by drinking the blood of the Ahlus-Sunnah. (Umar 2013a, 51)

Umar (2013b) describes the people of Syria as "sisters and daughters of the Sunnis," establishing a kinship and in-group through belief against the Shia. Umar emphasizes the bloodthirsty nature of the Shia by alleging that they killed important figures in Islam. Muhammad Farhad Janbaz (2013b) continues this theme of Sunni victimization by attributing the Assad regime's cruelty to him being Shia: "The foundational development of this sect is clear to all Muslims of the *Ummah*. They have displayed their evil character, having conspired to harm the nation's majority through different means" (42). Shahid Zui (2013a) recites a litany of abuses against the Assad regime, connecting its treachery against Sunni Muslims to Syria's foreign policy: "Due to committing atrocities against Muslims, they acquired Russia's protection and were acceptable to America and other Western countries. At one time, they obtained the gift of military aid from Russia and financial aid from the West" (9–10). Zui (2013a) holds the United States responsible for Syria's chaos: "American and European intelligence agencies attempted to create obstacles on the path of finding a political solution to the Syrian problem to smooth the way for sectarianism in the Middle East by subjugating Muslims through Bashar al-Assad's forces, Iran's support for the Syrian government, and placing Hezbollah in the middle" (Zui 2013a, 11). In each case, Taliban authors define the out-group as Shia Muslims as a cognitive shortcut to disparage the Assad government as an enemy.

Turkey

Taliban authors view Turkey through an ally image consistent with themes of sympathy for the Ottoman caliphate in Taliban discourse. An anonymous article in *Al-Somood* entreats Turkey as a Muslim power to remove its forces from Afghanistan: "Turkish forces sent a maritime convoy to Afghanistan after Turkey consented to France's task of leading Kabul's forces from November 2009. We say to him [Rajab Tayyab Erdogan] that this action is very painful to all Afghans and all Muslims since Turkey has a meaningful and great historical status among all Muslims" ("Fī Intidhār 'Ghadhba 'Uthmānīya' Min Al-Raīs Al-Turkī" 2009). The article advises Erdogan to withdraw forces: "Announce the withdrawal of your forces from Afghanistan, calling for the ceasing of this tyrannical war and for the withdrawal of all enemy forces from it. And we do not say that you should order your forces to relinquish their weapons completely to the mujahideen of the Islamic Emirate since that would be above your power and outside of the secular beliefs of your army" ("Fī Intidhār 'Ghadhba 'Uthmānīya' Min Al-Raīs Al-Turkī" 2009). Similarly, Aimal Nuri (2013) has concluded that the Taliban released a Turkish engineer for Islamic solidarity: "This incident of freeing a Turkish engineer bears witness to the fact that the Taliban wants good relations with the world as a matter of practicality, especially with the Islamic world. In the future, it has the capacity to come forward as a strong political power" (29). Both examples highlight the extent to which an in-group formed through religion can influence perceptions of foreign policy.

Therefore, Taliban authors praise Erdogan's muscular foreign policy as a leading Muslim power. Momand (2013b) writes of the Turkish response to the 2010 Israeli–Palestinian war:

> It was only Turkey that showed determination in taking steps to fully help the blockaded Muslim brothers in Gaza. A caravan of ships containing aid departed from Istanbul for Gaza, but Israel's commandos violently attacked the caravan en route in which 79 Turkish individuals were martyred and 56 injured. The government of Turkey condemned the incident in harsh words and told Israel to accept its error, give the descendants of the martyrs compensation, and end the blockade of Gaza. (18)

Momand (2014b) views Erdogan as the leader of a resurgent Turkey unwilling to obey American and European powers:

> He took steps that definitely became the cause of anger for Israel, the Western world—especially the United States—and the reason for starting a conspiracy to end Tayyab Erdogan's government. For example, those mosques that were made into museums were made into mosques again and opened for prayers. The call of monotheism started resounding from minarets again, respect for Islamic practices started again, and women reached the parliament in those very veils whose ban he lifted. (22)

Momand (2014b) predicts that Erdogan's Islamist sympathies will be the reason for his downfall: "The foreign policy of the current government of Erdogan will never be acceptable to America, the Western world, and Israel, and they will definitely make a plan to end this government" (22). Similar to their own self-image vis-à-vis Afghanistan, Taliban authors relate to Erdogan as the leader of a nation-state that boldly defends Muslim interests.

The United States of America

Taliban authors have written dozens of texts against the United States as an imperialist power. Alā Shamāsina (2010) sought to dispel popular conceptions that President Obama's foreign policy would differ substantially from President Bush's: "American policy is not an individual policy that changes with a change of individual. Rather, it is a preparation from the American kitchen, and the thoughts preparing this American policy have not changed even though the face of the implementer of this policy has changed. Obama and Bush are two faces of a single coin" (51). Shamāsina (2010) contends that American foreign policy is a ploy to dupe Muslims: "It is not rational that change in foreign or American policy would be in the wellbeing of the *ummah* and Muslims since it is a colonizing policy that raises the Crusader for war and carries hatred in its depths" (51). Jaffer Hussein (2013) reinforces this image: "Obama's election, his staunch popularity among the Jewish lobbyists and his blanket support to Israel shows some of the real intentions behind his foreign policies. But the average American must question himself as to why America has gone on and invaded other countries, killed their people and spread bases all over the world. What is the need for all this when the leadership does not claim to possess a religious or an ideological agenda?" (36). Shamāsina and Hussein portray the United States as a nonsecular state, driven by Judeo-Christian interests to colonize other states.

Other Taliban authors ascribe varying geopolitical motives to American imperialism. One article in *Srak* suggests that the United States seeks to exploit Afghanistan's natural resources: "The Americans have discovered the heavy mineral lithium in Afghanistan and recognized that not only is lithium useful for computer systems, but also that the application of this metal can be greatly transformed into military weapons. . . . America's military campaign in Afghanistan is for the purpose of economic profit. Otherwise, it would not have sacrificed so many of its youth" ("Ma'ādin-e Afghānistān Dar Changāl-e Ghāratgarān" 2013). Another article extends the theme of American imperialism even to NATO allies, alleging on the basis of WikiLeaks files that officials have expressed concerns with increasing casualties in Afghanistan and will only occupy peaceful areas: "Washington's search to convince European allies to send more troops to the south [of Afghanistan] concluded with Bonn, and countries like Germany and France established themselves in peaceful areas" ("Sar Khordegi Wa Tars-e Hampaimānān-e Amrīkā az Jang-e Afghānistān" 2010). An unnamed Dari writer in *Srak* states that one of America's long-standing goals is to have "a presence in the region that would stand atop Russia, China,

Iran, and the Persian Gulf" ("Tala-e Afghānistān" 2010). This goal exacts a steep price:

> Now a big part of the country of Afghanistan is outside of the control of the Kabul government and foreign NATO forces and is established in the authority of the Taliban. Governmental corruption and bribery in Afghanistan are prevalent, and this problem increases daily despite the anger and exasperation of the people from American and other European member countries of NATO whose money and wealth is being wasted in Afghanistan on a failed and limitless military path. ("Tala-e Afghānistān" 2010)

This Dari writer acts as a bridge to the second type of Taliban image of the United States as a degenerate power, in which military leaders lack cultural sophistication and the support of the American public. Shihabuddin Ghaznavi (2008a) disparages the United States for its barbarity in justifying destruction in Afghanistan through the 9/11 attacks:

> America presented the 9/11 incident as a justification for its attack against Afghanistan, and in exchange for this incident, its airplanes started to bomb the people of Afghanistan in its initial days which led to the deaths of thousands of women, elderly, children, youth, the afflicted, and the injured. And that was their revenge such that blind bombing led to injuring almost sixteen thousand in addition to destroying their houses, razing their mosques, and burning their fields. These barbaric tyrannies continue to this very day, and there isn't a day that passes in which the Crusader forces do not undertake killing thousands of innocent civilians. (20)

Taliban authors see themselves as defenders of Islam and Afghanistan against American degeneracy. Hanif Hammad (2012) compares the moral poverty of American technology to the richness of Taliban courage through an Arabic proverb: "This reminds me of a famous Arabic saying, '*As-saif bi as-sā'id la as-sā'id bi as-saif.*' Meaning that the hand uses the sword, not the sword uses the hand! That behind the sword, the real power is that strong hand that moves it. If that hand is weak, then a strong sword is of no use. This reality bring us to the conclusion that to use weapons on the battlefield, it is necessary for a man to be brave and courageous" (17). Muhammad Adil (2012) also depicts the United States through symbols from Arabic and Islamic history, drawing an analogy with the Abrahamic prophet Moses: "The kind of Pharaonic behavior of the modern, world-leading America is just like the deranged elephant and Moses' Pharaoh who is not ready to accept its weakening grasp of leading the world like sand slipping from dirt. The brave and daring Taliban, having given this elephant an injuring wound of a strong and continuous resistance, is not leaving it capable of running from the battlefield safe and sound" (39). Adil (2012) likens America's technology to "sorcery" and the Taliban's power to the "white hand of Moses" to reinforce a cognitive shortcut of the United States as a contemporary Pharaoh (39).

For this reason, Taliban authors frequently write of the need for the United States to exit Afghanistan. An unattributed article in *Sharī'at* cautions readers against any strategic agreement that would place American forces in Afghanistan beyond its stated combat missions: "The Islamic Emirate will continue its armed struggle from the support of the free Afghan people with its full force against the provisions of this illegal agreement until all the occupying forces and their paid-off slave [Hamid Karzai] have been chased and removed from Afghanistan" ("Karzāī Aur Obāmā Kē Mā Bēn Istratijīk Muāhidē Par Dastakhat Par Imārat-e Islāmīya Kā Radd-e Amal" 2012). The Taliban warns: "This holy jihad will continue until the occupying forces are removed and are the reason for the deployment of forces against America and its allied forces" ("Karzāī Aur Obāmā Kē Mā Bēn Istratijīk Muāhidē Par Dastakhat Par Imārat-e Islāmīya Kā Radd-e Amal" 2012). One official statement declares: "If the United States wants lasting peace in Afghanistan and in the region, is serious about saving its nation from the presently heavy crisis, and practically wants to end this illegal and pointless war, then it should quickly take practical steps to evacuate its forces from Afghanistan" ("Amrīka Aur Karzaī Intizāmīya Kē Darmīan Honē Wālē Aman Muāhide Kē Mutalaq Imārat-e Islāmīya Kā Elāmīya" 2013).

According to some authors, America's degeneracy even threatens its own military. Mustafa Hāmid (2012a) asserts: "American soldiers are becoming the victim of mental problems and as a result, their blood pressures are rising such that the predominance of suicide among American soldiers is increasing. After soldiers dying from the bullets of the mujahideen, the second largest number [of casualties] is of those who die by committing suicide. This is all a result of the Taliban's bravery, righteous courage on the battlefields, and military skills" (27). Hāmid (2012a) blames American officers for killing their own soldiers: "Despite the length of the war and large number of non-official forces in the Taliban's control, there has been no living soldier in the Taliban's control aside from Bergdahl. Therefore, it is surmised that the American military is suspected of killing its own soldiers; those soldiers who have surrendered are bombed by air and freed from a life of imprisonment" (28). For this reason, Rahimi (2012) contends that American soldiers shun war altogether upon return to the United States: "When they return home, most American military officials leave duty or opt for rebellion and some restrict their lives to four walls. Some enter the NGO field and some take faith having accepted Islam after being impressed by the morals of the Taliban" (34). Habibullah Yusufzai (2014) contends that President Obama's planned departure of troops from Afghanistan in 2014 is proof of America's defeat, asking what America has accomplished after thirteen years of war: "In the end, its economy has gone out of control with trillions of dollars wasted, tens of thousands of American troops in the face of America's war-mongering policies have met with destruction, been injured, are afflicted with mental disturbances, and are enduring the miserable sorrows of war without a goal and without an end, who without a doubt had wishes, goals, and also the right to live in this world" (70). Hāmid, Rahimi, and Yusufzai all represent American leaders as playing with the lives of American soldiers for no clear gain, despite their superior military strength.

In contrast, the Muslims of America are portrayed as allies. Muhammad Zakariyya (2013) explains that politicians have consistently misled American Muslims:

> Just like George Bush Jnr before him, Barack Obama has gone to great lengths to try and convince the Muslim population of America that they are at war with "terrorism" and not Islam or the Muslims. Time and again, this lie has been exposed by the very actions of the American government, police and army. This time the NYPD (New York Police Department) has uncovered its true face by secretly labeling mosques as "terrorism groups." This means that anyone who prays in that mosque is fair game to be suspected as a terrorist and have surveillance on his every action. Additionally the NYPD has been trying to infiltrate the boards of mosques with their informants. No Muslim in America can feel safe from police persecution. (57–58)

Zakariyya (2013) appeals to American Muslims by equating the actions of a domestic police force trying to infiltrate the boards of mosques with their informants and an army deployed internationally at war. The only solution is migration to an Islamic country for jihad: "It's time to follow the likes of Imam Anwar al-Awlaki and Samir Khan (May Allah accept them both) in abandoning the nation that is at war with Islam and joining the Mujahideen" (57–58). The common goal becomes an international defense of Islam against an out-group of infidels.

• • •

Discourse analyses of images for nation-states in Taliban texts confirm the hypothesis that differing authors consistently represent Muslim populations as allies and all others as enemy, degenerate, and imperialist powers. This finding corresponds to the Taliban's formation of in-groups and out-groups along the dimension of religion, as shown in chapter 3. While there may be a one-to-one correspondence between Muslim allies as an in-group, there is a range of possible images for infidel out-groups. Hermann and Fischerkeller (1995) suggest possible reasons for this finding in arguing for more images beyond the traditional ally and enemy:

> Relying exclusively on the enemy image greatly limits the analytical and explanatory utility of cognitive perspectives. It forces us to describe all relationships as essentially threat-based security dilemmas among actors roughly comparable in capability. This reduces our ability to capture conceptually the variation neorealists and other theorists are introducing on both motivational and capability dimensions and leaves us unprepared to wrestle with the most important distinctions made by area specialists. (424)

To correct this ally–enemy binary, Hermann and Fischerkeller (1995) state: "What is necessary is a set of images that captures the most important aspects of differently perceived strategic situations and that suggests different strategic alternatives" (424). Images must be rooted in empirical data "to examine how actors see the power relationships and the motivation and cultural norms of other actors. These images of other actors affect expectations about behavior—such as whether force will be used, the presumed relevance of norms, and calculations about how best to achieve objectives" (448–449). Such an examination would better capture the range of relationships in a post–Cold War world while integrating cognitive psychology with foreign policy (Hermann and Fischerkeller 1995, 450).

The War on Terror introduces another need to refine image theory by interrogating the nation-state as the sole actor in the international system. Some have sought to include non-state actors within image theory (Risse 2002). The Taliban enjoyed de facto, if not de jure, status as a nation-state actor when it occupied institutions of power from 1996 to 2001 and remains a significant insurgent force, even if it is not comparable in capability to other nation-states, such as the United States. Taliban authors have portrayed the Muslim populations of Egypt, India, Mali, Pakistan, Palestine, Syria, and the United States as non-state allies in contrast to their state governments, just as the Taliban has presented itself as the voice of the *nation* against the official Afghan *state*. If the ally image leads to a strategy of collaboration toward a common cause between the subject and target actors (Hermann and Fischerkeller 1995), then law enforcement and intelligence officials can investigate which populations the Taliban seeks to recruit in these specific countries. The only state routinely viewed by Taliban authors as an ally is Turkey, due to its past as the seat of the Ottoman Caliphate and the perception that it protects Muslim interests in the Middle East. Policy makers should consider whether Turkey could be enlisted as a mediator between Taliban and NATO forces.

Hermann and Fischerkeller (1995) contend that the value of image theory is that images seem to be associated with different foreign policy strategies. Different strategies by image also seem to be borne out in Taliban discourse. For example, the Taliban has not attacked Burma, India, or Israel, since they are perceived as aggressive and threatening enemy actors. Even assuming that the Taliban had military capabilities to conduct distant attacks, texts about these countries do not call for direct attacks against them. Instead, Taliban authors have tried to contain their influence by building alliances, and the Internet has permitted the Taliban to position itself as a global Islamist movement in the virtual emirate. Notably, each of these countries has a Muslim minority living among a non-Muslim majority—Buddhists in Burma, Hindus in India, and Jews in Israel—enabling the Taliban to publicize pan-Muslim grievances. In contrast, Pakistan elicits an image of degeneracy as a disorganized and chaotic state, and Taliban authors have encouraged direct attacks by inciting mujahideen. It is unlikely that government actions against militancy in Pakistan will alter this image, since Taliban authors could continue portraying the military as attacking fellow Muslims. Instead, counterterrorist efforts should consider the involvement of religious scholars, since Taliban authors have expressed consternation at possible religious rulings against the mujahideen and vaunted the certification of the Taliban as a legitimate movement by Pakistan's Grand Mufti (chapter 1). Finally, Canada, England, France, and the United States are depicted as imperialist powers; Hermann and Fischerkeller (1995) suggest that the imperialist image is based on the perception of another state's superior military capabilities and that direct attacks are conducted against a client regime rather than the imperialist. This is indeed the case, as Canada, England, France, and the United States contribute soldiers to NATO in defense of the official Afghan government, which Taliban authors see as weak and beholden to foreign interests. It is unlikely that their images as imperialists will change given the persistent image of Russia as an imperialist force years after the Soviet withdrawal from Afghanistan.

Empirical analysis of Taliban images also exhibits the role of cultural variables in cognitive information processing. The most common types of cognitive processes include *in-group favoritism*, in which Taliban authors establish group solidarity with Sunni Muslims elsewhere, and *analogous reasoning*, as the history of colonialism is used to understand current circumstances. Both cognitive processes illuminate shared understandings and expectations in the virtual emirate. Taliban authors treat Sunni Muslims as a monolithic community that would be receptive to their messages, avoiding discussions of theological differences or variations in practice. Local approaches to Islam based on lived experience are ignored as the worldwide Muslim community is mobilized for political action. By the same token, foreign involvement in Afghanistan reignites the memory of colonialism against which the Taliban exhorts Muslims to act. Taliban authors therefore culturally work out the differences between right and wrong in the virtual emirate by treating religion as a marker of social identity on whose behalf the Taliban purports to speak. These themes persist across texts in different languages without significant linguistic differences, in contrast to themes from the previous chapters. I suspect that this relates to the theme of the nation-state under discussion. We have seen that the Taliban uses the Internet to attract followers beyond Afghanistan when discussing Mullah Omar's leadership for the worldwide *ummah*, enlisting Muslims against infidels everywhere, and inviting others for jihad. Taliban authors are cognizant of the differential distances traveled by Arabic, Dari, English, and Urdu, exploiting literary canons and geocultural themes of power for maximal aesthetic effect. In contrast, Taliban authors of texts on nation-states seem focused on ideological arguments to the exclusion of any aesthetic dimension and may not be drawing on intralinguistic elements of poetry, literary canonicity, and genre style that emphasize aesthetics. It is possible that readers feel emotionally moved upon reading about injustices against fellow Muslims abroad or the potential repetition of colonialism in Afghanistan, rendering any need for additional aesthetic flourish redundant. For this reason, empirical studies of image theory expand our understanding of how groups differentially deploy elements of culture in cognitive information processing, underscoring that culture remains a dynamic process of meaning-making in different situations.

Epilogue

IT HAS BEEN A TROPE of our times to call for changes in the narrative of terrorism. On February 18, 2015, President Obama (2015) concluded a summit on countering violent extremism, declaring:

> Just as those of us outside Muslim communities need to reject the terrorist narrative that the West and Islam are in conflict, or modern life and Islam are in conflict, I also believe that Muslim communities have a responsibility as well. Al Qaeda and ISIL do draw, selectively, from the Islamic texts. They do depend upon the misperception around the world that they speak in some fashion for people of the Muslim faith, that Islam is somehow inherently violent, that there is some sort of clash of civilizations.

President Obama (2015) called for a new collaboration based on a counternarrative: "When all of us, together, are doing our part to reject the narratives of violent extremists, when all of us are doing our part to be very clear about the fact that there are certain universal precepts and values that need to be respected in this interconnected world, that's the beginnings of a partnership." President Obama's frequent invocations to "narrative" within government policy, political science, and international relations circles invite a broader discussion into the general role of communication. I have shown from the first chapter that the Taliban challenges the ability of the West—essentially, the United States and NATO allies—to set the terms of this debate by using new media technologies to articulate cultural differences, despite living in an interconnected world. The framework of the virtual emirate highlights three critical themes that allow Taliban authors to claim to speak for Muslims: (1) the Internet acts as a virtual reality that transcends geographically restricted notions of space; (2) the Taliban routinely claims victories in its struggle for sovereignty to control the lives, deaths, and conditions of existence for others through the Islamic Emirate of Afghanistan; and (3) Taliban authors articulate the cultural distinction between good and evil in everyday life based on a militant, political interpretation of Shari'a law that collapses the boundary between private and public practice of religion. In this sense, the Taliban quest for legitimacy in virtual reality mirrors its quest in physical reality, a confirmation of the finding that Internet communication reproduces the sociopolitical relations of the physical world (Cook 2004). I have followed Taliban authors from their first forays into the Internet during the 1990s until today to reveal longitudinal transformations of identity; the Taliban now presents itself as a spokesperson for all Sunni Muslims compared to its origins as a local Afghan militia, positions itself as a non-state actor in the international system, and relates to others along the dimension of religion.

The power of narrative also invites a broader discussion into the seminal roles of language and discourse in framing messages. Based on Sheldon Pollock's (2006) formulation of cosmopolitan and vernacular languages as a way "to relate literary-language choice or change and narratives of literary history to their most salient conditions, the acquisition and maintenance of social and political power" (Pollock 1995), I have attempted to prove with empirical data that Taliban authors possess a taxonomy of languages based on their perceived distances of travel: English and Arabic are cosmopolitan languages used around the world, Urdu is parochial to the Muslims of South Asia, and Dari is vernacular for Afghan Muslims. We observed differences in language use, such that Taliban authors issue Mullah Omar's orders in Urdu to reach South Asian Muslims (chapter 2), and the conception of Arabic as a language of the Muslim world (chapter 3). This differential language use exemplifies one definition of linguistic ideology as "sets of beliefs about language articulated by users as a rationalization or justification of perceived language structure and use" (Silverstein 1979, 193). Taliban authors frame arguments through the literary cultures of these languages with their readers, from writing in strict rhymed prose for Arabic audiences to situating the jihadist travelogue within the Persian genre of travel writing to quoting legendary poets in Arabic, Persian, and Urdu for maximal effect. Taliban authors commemorate martyrdom through the Quran in Arabic texts but opt for the high-literature poets of Hafez and Sa'di in Persian. Taliban authors create new forms of culture that Pollock (2000) has identified through the literary processes of studying canonical authors, observing language discipline, and adhering to rules for rhetoric. Any initiative toward changing militant narratives—which seem to be the ideological components of a West–Islam conflict, mentioned by President Obama—must also include an affective component that incorporates aesthetic norms inherent to each language. Taliban narratives attend to style as well as substance, and so should counternarratives. The appeal for counternarratives is likely to endure, as President Obama and Afghan president Ashraf Ghani have announced an extension of the American troop presence in Afghanistan through the end of 2015 (White House 2015).

This raises the question of which types of counternarratives are likely to be successful. Governments may be tempted to disrupt or block militant websites altogether, but this has only motivated the Taliban and other groups to find other online avenues. Therefore, ongoing censorship is unlikely to succeed as a response. Instead, the fusion of general theories from political science and counterterrorism studies with empirical data from this book can offer concrete directions. The Center on Global Counterterrorism Cooperation (Shetret 2011) recommends three strategies that resonate with my findings: (1) weaken cult personalities, (2) challenge extremist doctrine, and (3) dispel the glory of the militant lifestyle. The late Mullah Omar qualifies as a prominent figure with a cult personality that the United States and other allies could counter, but his disappearance from public view and delayed death announcement have only contributed to his mythical status among followers. One possible countermessaging strategy could create distance between the new supreme leader Mullah Akhtar Mansoor and his district commanders, publicizing Taliban brutalities and atrocities at the district level. District commanders do not enjoy the same reputation for morality and piety as the late Mullah Omar or Mullah Mansoor and may be easier targets. To challenge extremist doctrine, countermessaging efforts should mine the same literary sources—whether religious texts such as the Quran and Hadith or secular literature such as poetry—that the Taliban uses to buttress its arguments. Interpretations of religious texts from South Asian scholars trained in Deoband or Hanafi theologies or from global centers of Sunni learning, such as Al-Azhar University in Egypt, can be disseminated to promote nonviolent interpretations of jihad. To dispel the glory of the militant lifestyle, counterpropaganda efforts can publicize narratives of the growing number of former Taliban officials who have abdicated violence. These individuals possess a type of legitimacy among potential recruits that can only be accrued through prior involvement in militancy. Their reasons for leaving the Taliban and responses to integrating within civil society can encourage others. A full countermessaging strategy should attend to aesthetics and cutting-edge technologies just as the Taliban's multimedia exploits high-quality, slick production.

Despite writing in different languages, Taliban authors disseminate a unified message that crystallizes a clear cultural system. Let us review the definitions of *culture* and *psychology* in this book. Culture is "the range of shared understandings and expectations that are embedded in individual psychology, societal institutions, and public practices" (Renshon and Duckitt 1997, 233). Psychology "refers broadly to mental functions, such as perceiving, categorizing, reasoning, remembering, feeling, wanting, choosing, and communicating" (Shweder 1999, 66). Discourse analysis allows us to study shifts in identity, new functions of language, and the circulation of ideas in society to observe power dynamics in which certain shared understandings and expectations dominate others (Foucault 1991). Examining individual texts for thematic convergences and divergences provides a ground-level look at the formation of shared understandings and expectations. This approach starts from the individual as a unit of analysis rather than the group. Anthropologists have shown that the group approach to culture in psychology and psychiatry too often assumes that individuals can be reduced to their race, ethnicity, language, or customs (Guarnaccia and Rodriguez 1996; Carpenter-Song, Schwallie, and Longhofer 2007). As I showed in chapter 4, the notion that the Taliban espouses jihad because its members are ethnic Pashtuns or observant Deobandi Sunni Muslims does not explain why most members of these groups shun violence.

The cultural critic Homi Bhabha (1994) reminds us, "It is in the emergence of the interstices—the overlap and displacement of domains of difference—that the intersubjective and collective experiences of *nationness*, community interest, or cultural value are negotiated. How are subjects formed 'in-between', or in excess of, the sum of the 'parts' of difference (usually intoned as race/class/gender, etc.)?" (2). The Taliban attempts to form cultural subjects at the interstices of different domains, such as language group, ethnicity, or national citizenship, cautioning us against equating culture with any group affiliation. Psychiatrists, psychologists, and social scientists studying the Internet's role in spreading religious justifications of violence (Bhui and Ibrahim 2013) may benefit from the methods in this book that explore such cultural interstices. These methods include the following:

1. Examining how language is used as a common medium for social communication
2. Showing how leaders transmit shared meanings and expectations to followers
3. Observing in-group/out-group construction through self-derived categories
4. Investigating organizational missions shared among in-group members
5. Tracing relationships between out-group stereotypes and images of world actors

These methods can be applied to analyze other militant organizations that mobilize religious meanings to incite violence, such as Al Qaeda and the Islamic State. The 9/11 Review Commission (2015) convened by the U.S. Congress has recommended that law enforcement and intelligence agencies counter violent extremism by both capitalizing on state-of-the-art technology to disrupt social media use among militants and integrating linguists within intelligence gathering and analysis. As engaged critics, behavioral and social scientists can inform this effort by creating and contesting such methods to advance social theory in culture and psychology across academic disciplines while improving government policy in the War on Terror (Aggarwal 2015). Military and law enforcement officials preserve order in the world as it currently exists, while social and behavioral scientists seek to understand *why the world exists in this form*, explaining a fundamental difference in their approaches to problem solving. Fields such as cultural psychiatry and cultural psychology can document the range of human experiences while reminding us of our common humanity, no matter how unrecognizable in these times of terror.

NOTES

1. CHANNELS OF COMMUNICATION IN THE VIRTUAL EMIRATE

1 The Pakistani scholar Maulana Fazhl Muhammad flaunted the Taliban's connections to Pakistani madrasas on the Taliban's website before 2001: "The fact that the Taliban are the students of the Madresa [*sic*] is sufficient for the Muslims to place their trust in them. The Governor of Jalalabad is a graduate of Darul-Uloom Haqqania, Akora Katak. The judge of the high court is a graduate of another Madressa in Peshawar. The judge of Khost is an 'Alim [scholar] graduated from a Madressa [*sic*]. The representative of the Taliban for the U.N. Maulana 'Abdul Hakim is a graduate of Binnori Town, Karachi and a class mate [*sic*] of mine. The Afghan Ambassador in Islamabad Mufti Ma'soom is a graduate of Darul 'Uloom, Karachi and has attained his course of Mufti by Mufti Rasheed Ahmad Saheb. The Ambassador in Karachi is a graduate of Jamia Hammadiyah" (Muhammad 1998).

2 At that time, Al-Rasheed Trust was located in Pakistan: "Al-Rasheed Trust, Kitab Ghar, opposite Dar-ul-ifta-e-Wal Irshad, Nazimabad 4, Karachi. Tel: (0092) (021) 6683301 Fax: (0092) (021) 623666 Email: *dharb@cyber.net.pk*" (Islamic Emirate of Afghanistan 1998).

3 The dispute between Afghanistan and Pakistan over Pashtun territories along their borders has been long-standing. The Durand Line was negotiated in 1893 by British official Sir Mortimer Durand and amir (king) of Afghanistan Abdur Rahman to prevent Russian encroachment (Omrani 2009). Afghanistan has argued that the Durand Line only clarified political spheres of influence, whereas Pakistan has argued that the Durand Line is a legal border (Omrani and Ledwidge 2009). Since the 1970s, Pakistan has attempted to increase its influence in Afghanistan's domestic affairs to prevent Indians and Afghans from co-conspiring to destabilize Pakistan (Haqqani 2008). A declassified letter

to President Richard Nixon from Vice President Spiro Agnew after a private meeting with Afghanistan's King Zahir Shah (1933–1973) noted: "The principal worry on the King's mind was the long-standing Pushtoonistan dispute with Pakistan.... The King said that even though the dispute with Pakistan was an emotional one, there was no evidence of any breakout of hostilities; and he felt that so long as both sides continue to suggest new bases of settlement the matter could be eventually resolved" (Agnew 1970).

4 Not all Taliban members supported bin Laden. American embassy reports indicate that up to 80 percent of the Taliban's leadership opposed bin Laden's presence in Afghanistan but that he was protected by Mullah Omar (U.S. Embassy [Islamabad] 1998a). Taliban leaders secretly met with American government officials in 2000 to resolve the bin Laden issue, with one Taliban leader even praising the United States for help with jihad against the Soviet Union (U.S. Embassy [Islamabad] 2000). Some strongly supported Osama bin Laden, as in this article from Maulana Zahid-ur-Rashidy (1998): "I happened to interview with Osama in Afghanistan, the detail of which was published in national papers. Time is ripe to review objectively the dispute between America and Osama, which compelled the Billionaire Merchant Prince to take Klashnikov against America and lead a camp life. Actually Osama has every right to take arms against America, as 'the Media' and all the means of 'Lobbying' are available to his opponents, who are casting aspersions upon him. This is his last resort."

5 The neo-Taliban's promotion of opium contrasts plainly with the original Taliban's edicts. Mullah Omar ordered local councils to ban poppy cultivation in 2000 or risk public execution, leading to a 99 percent reduction of domestic poppy cultivation in 2001 and a 65 percent reduction in global supplies (Farrell and Thorne 2005). However, the Taliban may have restricted poppy cultivation more for international recognition than for religious reasons; American and Afghan officials suspect that Mullah Omar has opium stockpiles and that he taxes opium farmers for revenue (Peters 2009). American officials estimate that the neo-Taliban has raised $70–$400 million in opium cultivation, taxes on farmers, and trade fees, as well as $100 million in foreign donations, frustrating counterterrorism efforts (Schmitt 2009). The Taliban disputes these allegations, contending that opium production has been an American ploy "to numb the power of the Islamic World, defeat its forces, and subsequently smooth the way to its conquest" since the 1980s, when the CIA established

labs and dispensaries to raise revenue for war against the Soviet Union (Ghaznavi 2008b, 44).

6 The Afghan neo-Taliban and the Pakistani Taliban (TTP) are separate entities despite sharing the "Taliban" name and the Afghan Taliban's leadership residing in Pakistan. Afghans have only haltingly cooperated with Pakistani fighters, who are seen as foreigners (Farrell and Giustozzi 2013). The TTP emerged in 2004 after Pashtun militant groups in Pakistan's Federally Administered Tribal Areas (FATA) negotiated separate peace treaties with the Pakistani government, but by 2007, the TTP had at least five thousand militants from leader Baitullah Mehsud's faction alone, as well as a forty-member council with representatives from all districts in FATA and Khyber-Pakhtunkhwa province (Abbas 2008). The TTP has collected revenue by taxing locals, charging truck drivers transit fees to smuggle goods, imposing stiff fines for violations of Shari'a law, promoting criminal activities throughout Pakistan, and housing Al Qaeda militants for fees (Acharya, Bukhari, and Sulaiman 2009). The number of TTP militants is estimated at 34,000–45,000 people of all nationalities (including Arabs and Central Asians), though new groups join and old groups leave, making statistical accounts unreliable (Qazi 2011). The TTP has frequently collaborated with the Punjabi Taliban—Sunni Islamist militants of Punjabi ethnicity who received state support to wage insurgency in Indian Kashmir in the 1990s but have now attacked Pakistan's security institutions after Pakistan abandoned support for the Afghan Taliban in 2001 (Abbas 2009).

2. MULLAH OMAR'S LEADERSHIP IN THE VIRTUAL EMIRATE

1 I use the conventional "Mullah" throughout this book, but retain "Mulla" in direct quotes.

2 This context of his dreams is crucial to understand. Interviews with one of Mullah Omar's physicians suggest that his dreams may represent seizures due to brain damage following a 1989 battle against the Soviets (Lamb 2001).

3 Kinnvall and Nesbitt-Larking do not acknowledge the source of the phrase "laws of the Father," but their references to the language of the Quran and Shari'a law conjure Lacan's phrase "the name of the Father," an interpretation of Freud's Oedipal complex through structural linguistics: "It is in the *name of*

the father that we must recognize the basis of the symbolic function which, since the dawn of historical time, has identified his person with the figure of the law" (Lacan 2006, 230). Lacan understood Freud's father figure as a symbol of the infant's socialization and acculturation mediated through language acquisition.

4 The "mirror-hungry" leader continuously requires praise: "These individuals, whose basic psychological constellation is the grandiose self, hunger for confirming and admiring responses to counteract their inner sense of worthlessness and lack of self-esteem. To nourish their famished self, they are compelled to display themselves in order to evoke the attention of others. No matter how positive the response, they cannot be satisfied, but continue seeking new audiences from whom to elicit the attention and recognition they crave" (Post 1986, 678–679).

The "ideal-hungry" follower continuously requires inspiration: "These individuals can experience themselves as worthwhile only so long as they can relate to individuals whom they can admire for their prestige, power, beauty, intelligence, or moral stature. They forever search for such idealized figures. Again, the inner void cannot be filled. Inevitably, the ideal-hungry finds that his God is merely human, that his hero has feet of clay. Disappointed by discovery of defects in his previously idealized object, he casts him aside and searches for a new hero, to whom he attaches himself in the hope that he will not be disappointed again" (Post 1986, 679).

5 Reparative leadership improves society: "By contrast, charismatic leader-follower relationships can also catalyze a reshaping of society in a highly positive and creative fashion" (Post 1986, 686).

6 The international worldview of the Taliban is the focus of chapter 5. This chapter presents Mullah Omar's international views.

7 Others have also claimed the title of Commander of the Faithful. On June 29, 2014, Abu Bakr al-Baghdadi (b. 1971), the leader of the Islamic State, proclaimed a caliphate in territories under his control in Syria and Iraq. The Twitter account of ISIS—@ISIS_Media_Hub—made the announcement: "Press secretary of ISIL al-Adnani announces creation of caliphate dawlah.ml/690QFQ."

Historically, there can be only one Commander of the Faithful for the Muslim community worldwide. For this reason, I searched ISIS and Taliban publications after the announcement to determine whether there was a Taliban response to this possible challenge to Mullah Omar. In fact, the domain name "dawlah.ml" links to a website known as the Kavkaz Center, which declares itself to be "a Chechen internet agency which is independent, international and Islamic. . . . The Kavkaz Centre agency makes it its mission to report real events in Ichkeria under conditions of a total information embargo and to disseminate to the world community the truth about the war, the war crimes, the evidence of genocide on the part of the state aggressor against the entire population and the positions of the side defending against aggression—the Chechen mujahedeen" (Kavkaz Center 2014). The website hosts content in English, Arabic, Turkish, and Russian. The ISIS tweet also linked to Mullah Omar's statement for Eid-ul-Fitr on July 27, 2014: "IEA. Mullah Omar addresses Muslims on eve of Eid-ul-Fitr dawlah.ml/6Rb3By."

One clue to this situation may come from Taliban texts. The Islamic State emerged from Abu Musab al-Zarqawi's (1966–2006) establishment of the Jamaat Al-Tawhid Wa-l-Jihad in 1999, also known as Al Qaeda in Iraq from 2004 to 2006, when it battled American forces in Iraq (Zelin 2014). The first issue of the Arabic periodical *Al-Somood* features Mullah Omar's eulogy of al-Zarqawi: "We therefore consider the mujahid brother Abu Musab al-Zarqawi a hero among heroes of the Muslim community, and likewise a warrior among warriors of the Islamic Emirate, and we are proud of his jihad against the Crusader forces" (Islamic Emirate of Afghanistan 2006b, 4). Eight years later, Taliban writer Hassan Momand (2014a) suggested that the mujahideen have so definitively defeated American and NATO forces in Iraq and Afghanistan that President Obama only wants to use air strikes, not ground troops, against ISIS: "Now those days have come that America and its allies will not be able to do anything despite their drones. The youth of the Islamic world will advance toward the Islamic State. Financial support from average Muslims will make the Islamic State stronger. The wrong suspicions and doubts among the Muslims related to the Islamic State will slowly end. Beyond the present situation, the Islamic State will become dangerous for the United States and Western powers" (8). However, some Taliban authors have openly questioned al-Baghdadi's allegiances. Shahid Zui (2014) states: "Al-Baghdadi has recalled the leader of Afghanistan's Taliban and Islamic Emirate Mullah Omar Mujahid as good hearted and has demonstrated his devot[ion] to him. But he only mentions Sheikh Osama [bin Laden], respects only him, and presents him with full tribute. For this reason, he has repudiated [Ayman] Al-Zawahiri's order" (11). These texts suggest that Taliban authors are trying to work out their relationship with the Islamic State publicly, though much remains uncertain. On June 4, 2015, the *New York Times* reported that the Taliban has had to fight an insurgency of its own against militants from the Islamic State in Afghanistan (Goldstein 2015).

8 The Taliban has a long history of naming its annual spring military operations after prominent figures and events in the classical Islamic tradition. The 2014 operation was named Khaibar, after the Prophet Muhammad's victory: "In the seventh *hijrī* year [seven years after the Prophet's migration from Mecca to Medina in 622] the war Khaibar took place against the enemies of Islam and as a result, formidable forts of the enemy and many of their large centers were vanquished. The infidels were completely expelled from the area and much wealth fell into the hands of the Muslims. This year, we have also taken a good omen from this name and we hope to God (may He be exalted) that our country will again be liberated from the hands of the infidels" ("Alamīya-e Shūrā-e Rahbarī-e Imārat-e Islāmī Dar Maurid-e Āghāz-e Amalīyat-e Jadīd Be Nām-e 'Khaibar'" 2014).

3. IDENTITY IN THE VIRTUAL EMIRATE

1 Nidal Hassan (b. 1970) is a former American Army psychiatrist who admitted to killing thirteen people and injuring dozens of others in the Fort Hood shooting on November 5, 2009. Tamerlan (1986–2013) and Johar (b. 1993) Tsarnaev killed three people and injured 264 others during the Boston Marathon on April 15, 2013.

2 The Arabic acronym *HA* stands for *Hayāhu Allah*, a blessing meaning "May God prolong his life."

3 Supposedly the sixth convert to Islam, he accompanied the Prophet Muhammad on military expeditions.

4 The abbreviation *RA* is an English acronym of the Arabic *Razī Allah ʿAn Hu*, "May God be pleased with him."

5 Sher (2014) presumes that his readership knows these figures. Information on Khalid bin Waleed can be found in chapter 2. Sultan Jalaluddin Khwarizm—the last ruler of the Khwarizm dynasty from 1220 to 1231—escaped the sacking of the Mongols; defeated them at the battle of Parwan, north of Kabul; razed Ucch, Sindh, and Gujarat in South Asia; and slaughtered the Christian population of Tblisi, Georgia, before being assassinated in Diyarbakir, Turkey. Noor-ud-Din Zangi (1118–1174) ruled Aleppo as part of his family's empire and massacred the Christian population of Antioch, a major military front

during the Crusades. Salahuddin Ayyubi (1137/1138–1193), better known by his Anglicized name Saladin, founded the Ayyubid dynasty over Egypt, Syria, much of Iraq and the Gulf, and parts of North Africa. Salahuddin defeated the Crusaders at the Battle of Hattin, restoring Palestine to Muslim rule after eighty-eight years of Christian control. Interestingly, Sher (2014) does not mention the military rivalry between Zangi and Ayyubi. Tipu Sultan (1750–1799) was the ruler of Mysore, who allied himself with the French against the British and other native armies in India. Imam Shamil (1797–1871) was the third Imam of the Caucus Emirate and revolted against Russian rule until his capture by the forces of Emperor Alexander II. Umar Mukhtar (1858–1931) led resistance forces against the Italian colonization of Libya. All of these individuals are renowned for their military prowess, and we can deduce that Sher is emphasizing a transnational martial culture.

6 The law reads:

> The following Act of Parliament received the assent of the President on 17th September, 1974, and is hereby published for general information:-
>
> Whereas it is expedient further to amend the Constitution of the Islamic Republic of Pakistan for the purposes hereinafter appearing;
>
> It is hereby enacted as follows:
>
> 1- Short title and commencement.
>
> (1) This Act may be called the CONSTITUTION (SECOND AMENDMENT) ACT, 1974
>
> (2) It shall come into force at once.
>
> 2- Amendment of Article 106 of the Constitution.
>
> In the Constitution of Islamic Republic of Pakistan, hereinafter referred to as the Constitution in Article 106, in clause (3) after the words "communities" the words and brackets "and persons of Quadiani group or the Lahori group (who call themselves 'Ahmadis')" shall be inserted.
>
> 3- Amendment of Article 260 of the Constitution.
>
> In the Constitution, in Article 260, after clause (2) the following new clause shall be added, namely—

(3) A person who does not believe in the absolute and unqualified finality of The Prophethood of MUHAMMAD (Peace be upon him), the last of the Prophets or claims to be a Prophet, in any sense of the word or of any description whatsoever, after MUHAMMAD (Peace be upon him), or recognizes such a claimant as a Prophet or religious reformer, is not a Muslim for the purposes of the Constitution or law (The Constitution of Pakistan 1974).

4. JIHAD IN THE VIRTUAL EMIRATE

1 The author uses the term *mujthamahu*, but that has no meaning. Assuming that this is a typographical error, I have translated this as "his society" from *mujtama'ahu*.

2 The Battle of Badr is one of the few battles mentioned in the Quran in chapter *Imran*: "When thou wentest forth at dawn from thy people to lodge the believers in their pitches for the battle—God is All-hearing, All-knowing—when two parties of you were about to lose heart, though God was their Protector—and in God let the believers put all their trust—and God most surely helped you at Badr, when you were utterly abject. So fear God, and haply you will be thankful" (Arberry [1955] 1996, 89). The Taliban invests this ancient battle with contemporary meaning by comparing its forces with the Prophet Muhammad's nascent community of believers.

3 An Urdu word inherited from Persian, *sarfarōsh* literally means "he who sells his head."

4 The first hemistich is in Arabic, and the Urdu is a literal translation of the Arabic. As mentioned earlier, the phrase refers to the Quranic expression *jihad fī sabīl Allah*, "jihad on Allah's path."

5. INTERNATIONAL RELATIONS IN THE VIRTUAL EMIRATE

1 ʿImād Al-Dīn Muhammad ibn Qāsim Al-Thaqafī (695–715) was an Arab general who conquered the territories of Sindh and Multan, now in Pakistan, to extend the Umayyad Caliphate. Mahmood Ghaznavi refers to Yamīn Ud-Dawla Abul-Qāsim Mahmūd ibn Sebüktegīn (971–1030), who conquered territories in eastern Iran, Afghanistan, and Pakistan. Abūl Muzaffar Muhi-ud-Dīn Muhammad bin Aurangzeb (1618–1707) was the sixth Mughal emperor. "Abdali" refers to Ahmad Shah Durrani, also known as Ahmad Khan Abdālī

(1722–1772), who is considered the founder of the Durrani Empire and of modern Afghanistan.

2 These are all revered figures in Islamic history. Umar is the second caliph mentioned in chapter 2. "Usman" refers to Uthman ibn Affan (577–656), who succeeded Umar as the third caliph. Ali ibn Abi Talib (600/601[?]–661) succeeded Usman as the fourth caliph for Sunnis, though Shia Muslims consider him to be the first leader after Muhammad, as he is Muhammad's cousin and son-in-law. This disagreement contributed to the Sunni–Shia schism. Hassan and Hussain are Ali's sons, who were both killed and are regarded by Shia, though not Sunni, as the second and third leaders of their historical community, respectively.

3 Abdullah Ibn Saba (c. seventh century) has been a polarizing figure in Islamic history. Some Shia deny his existence, and some Sunni claim that he introduced heretical ideas that were assimilated into Shia doctrine, such as the divinity of Ali (Anthony 2012). This Taliban author exploits his image to vilify Shia Muslims.

REFERENCES

Al-Somood. 2009a. [Untitled]. 3 (34): 3.

Abbas, Hassan. 2008. "A Profile of Tehrik-i-Taliban Pakistan." *CTC Sentinel* 1 (2): 1–4.

——. 2009. "Defining the Punjabi Taliban Network." *CTC Sentinel* 2 (4): 1–4.

Abdullah, Abu. 2013. "Qaum-parastī—?" [Nationalism—?] (Urdu). *Sharī'at* 2 (18): 25–26.

"Abtāl-Nā" [Our Heroes] (Arabic). 2006. *Al-Somood* 1 (4): 29.

Acharya, Arabinda, Syed Adnan Ali Shah Bukhari, and Sadia Sulaiman. 2009. "Making Money in the Mayhem: Funding Taliban Insurrection in the Tribal Areas of Pakistan." *Studies in Conflict and Terrorism* 32 (2): 95–108.

Adil, Muhammad. 2012. "Amrīkī Ālambardārī Ko Afghānī Dhachkā" [An Afghan Blow to American World Leading] (Urdu). *Sharī'at* 1 (4): 39–41.

Adler, Emanuel. 1997. "Imagined (Security) Communities: Cognitive Regions in International Relations." *Millennium: Journal of International Studies* 26 (2): 249–277.

Afghan, Walid. 2012. "Māhnāma-e *Sharī'at* Sach Kā Haqīqī Tarjumān" [The Monthly *Sharī'at* Is the Real Interpreter of the Truth] (Urdu). *Sharī'at* 1 (1): 2.

——. 2013a. "Morsī Hukūmat Kā Saqūt" [The Fall of the Morsi Government] (Urdu). *Sharī'at* 2 (11): 14–17.

——. 2013b. "Tālibe Par Tashaddud Aur Hindūstānī Khawātīn Kī Hālat Zār" [Violence Against a Female Student and the Plight of Indian Women] (Urdu). *Sharī'at* 1 (11): 6–7.

Aggarwal, Neil Krishan. 2008. "Muhammad Iqbal's Representations of Ram and Nanak." *Sikh Formations* 4 (2): 133–142.

——. 2010. "How Are Suicide Bombers Analysed in Mental Health Discourse? A Critical Anthropological Reading." *Asian Journal of Social Science* 38 (3): 379–393.

——. 2011. "Intersubjectivity, Transference, and the Cultural Third." *Contemporary Psychoanalysis* 47 (2): 204–223.

——. 2012. "Hybridity and Intersubjectivity in the Clinical Encounter: Impact on the Cultural Formulation." *Transcultural Psychiatry* 49 (1): 121–139.

——. 2015. *Mental Health in the War on Terror: Culture, Science, and Statecraft*. New York: Columbia University Press.

Agnew, Spiro. 1970. "Private Conversation with the King of Afghanistan." National Security Archive. January 21. Available at: http://www2.gwu.edu/~nsarchiv/NSAEBB/NSAEBB59/.

Ahmad, Abu. 2011. "Mashhad Min Bashā'a Al-Quwwāt Al-Fransīya Fī Wilāya Kābīsā" [A Scene of Among the Ugliness of French Forces in Kapisa Province] (Arabic). *Al-Somood* 6 (63): 12–13.

Ahmad, Aziz. 1964. *Studies in Islamic Culture in the Indian Environment*. Oxford: Oxford University Press.

Ahmad, Hafiz Saeed. 2012. "*Shuhadā-e Millat*" [Martyrs of the Nation]. *Sharī'at* 1 (2): 27–28.

Ahmad, Ishtiaq. 2010. "The U.S. Af-Pak Strategy: Challenges and Opportunities for Pakistan." *Asian Affairs: An American Review* 37 (4): 191–209.

Ahmad, Mirza Ghulam. 2010. *The Truth Revealed*. Tilford: Islam International Publications.

Ahmad, Mustafa. 2013. "The Awakening of the Ummah." *Azan: A Call to Jihad* 1 (2): 29–32.

Ahmadee, Saifullah. 2013. "Jihad." *Islamic Emirate of Afghanistan*. Available at: http://shahamat-english.com/english/index.php/articles/29453-jihad, accessed February 7, 2014.

Ahmadi, Wali. 2008. *Modern Persian Literature in Afghanistan: Anomalous Visions of History and Form*. New York: Routledge.

Ahrar, Saiful Adil. 2012a. "'Ālam-e Islām Khusūsan Mashriq Wusta Kī Nagofta Be Sūrat-e Hāl, Asbāb Aur Sadbāb" [The Unsaid, with Respect to the Condition, Causes, and Address, of the Islamic World and Especially the Middle East] (Urdu). *Sharī'at* 1 (2): 39–42.

——. 2012b. "Amrīka Kē Sāth Istratījak Muāhidē Kē Awāqib: Frāns Kē Sadāratī Intikhābāt Men Sarkozy Kī Shikast Kē Asbāb" [The Consequences of Strategic Agreement with America: The Reasons for Sarkozy's Defeat in France's Presidential Elections] (Urdu). *Sharī'at* 1 (3): 31–33.

——. 2012c. "Sarzamīn-e Afghānistān Par Britānvī Shahzāde Prins Herī Kā Fidāī Hamlon Se Pur Tapak Istiqbāl" [British Prince Harry's Tumultuous Reception in the Land of Afghanistan with Martyrdom Attacks] (Urdu). *Sharī'at* 1 (7): 19–21.

Akcali, Pinar. 1998. "Islam as a 'Common Bond' in Central Asia: Islamic Renaissance Party and the Afghan Mujahidin." *Central Asian Survey* 17 (2): 267–284.

Akhund, Mullah Abdur Rahman. 2012. "Al Badr Aur Al Fārūq" [Al Badr and Al Faruq] (Urdu). *Sharī'at* 1 (4): 4–5.

Alam, Muzaffar, and Sanjay Subrahmanyam. 2007. *Indo-Persian Travels in the Age of Discoveries, 1400–1800*. Cambridge: Cambridge University Press.

"Alamīya-e Shūrā-e Rahbarī-e Imārat-e Islāmī Dar Maurid-e Āghāz-e Amalīyat-e Jadīd Be Nām-e 'Khaibar'" [Announcement of the Leadership Council of the Islamic Emirate on the Subject of the Beginning of New Operations Named "Khaibar"] (Dari). 2014. *Haqīqat* 1 (3): 49–50.

Alarcón, Renato, and Edward F. Foulks. 1995a. "Personality Disorders and Culture: Contemporary Clinical Views (Part A)." *Cultural Diversity and Mental Health* 1 (1): 3–17.

——. 1995b. "Personality Disorders and Culture: Contemporary Clinical Views (Part B)." *Cultural Diversity and Mental Health* 1 (1): 79–91.

Alexander, Michele G., Shana Levin, and P. J. Henry. 2005. "Image Theory, Social Identity, and Social Dominance: Structural Characteristics and Individual Motives Underlying International Images." *Political Psychology* 26 (1): 27–45.

"Al-Jihād Al-Afghānī Wa Masūlīya Al-'ālam Al-Islāmī" [The Afghan Jihad and the Responsibility of the Islamic World] (Arabic). 2009. *Al-Somood* 3 (34): 1–2.

"Al-Munāsara Al-Muqāwama Al-Falistīnīya Farīdha Dīnīya Wa Masūlīya Islāmīya" [Aiding the Palestinian Resistance is a Religious Duty and Islamic Responsibility] (Arabic). 2008. *Al-Somood* 3 (31): 3.

Al-Sharq Al-Awsat. 2002. "Al-Mulla Omar Li' 'Al-Sharq Al-Awsat' Fī Awwal Hādith Sahāfī B'ad Harb Afghānistān: M'arikatuna Ma' Al-Amīrkīīn Badat Alān Wa Nīrānunā Satasil Ila Al-Bayt Al-Abyad" (Mullah Omar to Al-Sharq Al-Awsat in His First Journalistic Interview After the War in Afghanistan: Our Battle with the Americans Has Started and Our Forces Will Reach the White House) (Arabic). May 17. Available at: http://www.aawsat.com/print.asp?did=103706&issueno=8571.

"*Al-Somood*: Tuhāwir Batl Al-Jihadīīn" [*Al-Somood*: Dialogue with the Hero of the Jihadis] (Arabic). 2006. *Al-Somood* 1 (6): 3–4.

American Psychiatric Association. 1994. *Diagnostic and Statistical Manual of Mental Disorders, Fourth Edition (DSM-IV)*. Arlington: American Psychiatric Press.

Amin, Tahir. 1984. "Afghan Resistance: Past, Present, and Future." *Asian Survey* 24 (4): 373–399.

"Amrīka Aur Karzaī Intizāmīya Kē Darmīan Honē Wālē Aman Muāhide Kē Mutalaq Imārat-e Islāmīya Kā Elāmīya" [The Islamic Emirate's Announcement Related to the Upcoming Peace Agreement Between the Karzai Administration and America] (Urdu). 2013. *Sharī'at* 2 (11): 3.

Anderson, Benedict. 1991. *Imagined Communities*. New York: Verso.

"An Exclusive Interview with the German Mujahid, Brother Abu Adam." 2013. *Azan: A Call to Jihad* 1 (3): 59–71.

"Angrez Istimār Se Mulkī Āzādī Kī 93vīn Sāl Mukammal Honē Par Imārat-e Islāmīya Kā Elamīya" [The Islamic Emirate's Statement on the Completion of the 93rd Year of National Freedom from English Colonization] (Urdu). 2012. *Sharī'at* 1 (6): 4.

Anthony, Sean. 2012. *The Caliph and the Heretic: Ibn Saba' and the Origins of Shī'ism*. Leiden, Netherlands: Brill.

Anwar, Ikrimah. 2013a. "The Life After Death: The Grave." *Azan: A Call to Jihad* 1 (4): 9–16.

——. 2013b. "On U-Turns and the Pakistani Army Doctrine." *Azan: A Call to Jihad* 1 (1): 30–37.

Anwar, Ikrimah, and Maulana Muawiya Hussaini. 2013. "Let's Understand 'Suicide Bombing.'" *Azan: A Call to Jihad* 1 (2): 21–28.

Arberry, A. J. (1955) 1996. *The Koran Interpreted: A Translation*. Reprint, New York: Touchstone.

Aretxaga, Begoña. 2003. "Maddening States." *Annual Review of Anthropology* 32:393–410.

Asad, Talal. 1993. *Genealogies of Religion: Discipline and Reasons of Power in Christianity and Islam*. Baltimore: Johns Hopkins University Press.

Asani, Ali S. 2002. "Pluralism, Intolerance, and the Qur'an." *American Scholar* 71 (1): 52–60.

Asim, Ilyas. 2011. "Amāj-e Akhīr-e 'Amalīyāt-e Mujāhidīn Nishāndehande-e Kodām Rūhīye Ast?" [The Recent Target of Mujahideen Operations Is Indicative of Which Mentality?] (Dari). *Srak* 6 (62): 27–28.

Asim, Muhammad Isma'īl. 2010. "Hasht-e Saur: Nihāndehande-e Azādmunshī-e Afghānhā" [The Saur Revolution: An Indication of the Liberal-Mindedness of the Afghans] (Dari). *Srak* 5 (56): 9–10.

"Ātesh Dar Khirman-e Farhang" [Fire on the Heap of Culture] (Dari). 2013. *Srak* 8 (75): 16.

Atran, Scott. 2009. "To Beat Al Qaeda, Look to the East." *New York Times*, December 12, p. WK11.

——. 2010. "Turning the Taliban Against Al Qaeda." *New York Times*, October 26, p. A29.

Atwan, Abdel Bari. 2006. *The Secret History of Al Qaeda*. Berkeley: University of California Press.

Badakhshani, Saeed. 2012. "Anānīke Jasad-e Rasūl Allah Rā Be-Saraqat Borde Natawānastand Bar Shaksīyat-e U Ham Lakha Wāred Karde Namītawānand" [Those Who Have Not Been Able to Abduct the Body of the Prophet of Allah Cannot Also Stain His Personality] (Dari). *Srak* 7 (71): 74–76.

——. 2014. "Yahūd Dushman-e Dīrīne-e Musulmānān" [Jews, the Long-Standing Enemies of Muslims] (Dari). *Haqīqat* 1 (1): 8–9.

Badakhshi, Nur Allah. 2008. "Al-Hujūm Al-Amrīkī Ala Afghānistān: Munāqidh Al-Jamī' Al-Mi'āīīr Al-'ālamīya Wa Al-Qawānīn Al-Duwalīya" [The American Attack Against Afghanistan: Against All Worldly Standards and International Laws] (Arabic). *Al-Somood* 3 (27): 40–43.

Balochi, Saadullah. 2011. "Ay Ahl-e Maghrib! Kyā Tumhen Is Behbūde Tehzīb Par Nāz Thā" [Oh People of the West! Were You Proud of This "Improved" Civilization?] (Urdu). *Sharī'at* 1 (1): 39.

——. 2012. "Hum Kyōn Rāh-e Jihād Par Nikle" [Why Have We Set Out on the Path of Jihad?] (Urdu). *Sharī'at* 1 (4): 31–34.

Banuazizi, Ali, and Myron Weiner. 1986. Introduction to *The State, Religion, and Ethnic Politics: Afghanistan, Iran, and Pakistan*, 1–20. Edited by Ali Banuazizi and Myron Weiner. Syracuse, N.Y.: Syracuse University Press.

Barfield, Thomas. 2005. "An Islamic State Is a State Run by Good Muslims: Religion as a Way of Life and Not an Ideology in Afghanistan." In *Remaking Muslim Politics: Pluralism, Contestation, Democratization*, edited by Robert W. Hefner, 213–239. Princeton, N.J.: Princeton University Press.

——. 2010. *Afghanistan: A Cultural and Political History*. Princeton, N.J.: Princeton University Press.

"Barmī Musalmānon Kē Khūnī Sānhe Kē Bāre Men Imārat-e Islāmīya Kā Elamīya" [The Islamic Emirate's Statement About the Bloody Tragedy of Burmese Muslims] (Urdu). 2012. *Sharī'at* 1 (5): 4.

Bartholet, Jeffrey. 2001. "Inside the Mullah's Mind." *Newsweek*, October 1, p. 30.

Bartocci, Goffredo, and Roland Littlewood. 2004. "Modern Techniques of the Supernatural: A Syncretism Between Miraculous Healing and the Mass Media." *Social Theory and Health* 2 (1): 18–28.

Bawadi, Ahmad. 2009. "Mā Hīya Al-Shubhāt Al-Latī Afadhat Ila Al-Qu'ūd 'an Al-Jihād?!! Al-Halqa Al-Thāniya" [What Are the Suspicions That Have Led to the Failure of Jihad?!! The Second Part] (Arabic). *Al-Somood* 4 (39): 40–43.

"Bayān Al-Shūra Al-Qayādī L-il Imāra Al-Islāmīya Bi-Munāsaba Istishhād Al-Mujāhid Al-Kabīr Al-Shaikh Al-Shahīd Usāma ibn-e Lādin—Rahama-hu Allah Ta'āla" [Statement of the Leadership Council of the Islamic Emirate Regarding the Martyrdom of the Great Mujahid Shaikh Martyr Usama bin Laden—May God (May He Be Exalted) Have Mercy upon Him] (Arabic). 2011. *Al-Somood* 5 (60): 2.

"Bayān Imāra Afghānistān Al-Islāmīya Bi-Munāsaba Dhikra Al-'Asharūn Li-Insihāb Al-Quwwāt Al-Sofitīya Min Afghānistān" [Statement of the Islamic Emirate of Afghanistan in Relation to the Twentieth Anniversary of the Retreat of Soviet Forces from Afghanistan] (Arabic). 2009. *Al-Somood* 3 (33): 4.

"Bayān Imāra Afghānistān Al-Islāmīya Bi-Munāsaba Yaum Ihtilāl Afghānistān Min Qabla Al-Quwwāt Al-Rūsīya" [Statement of the Islamic Emirate of Afghanistan in Relation to the Day of the Occupation of Afghanistan by Russian Forces] (Arabic). 2008. *Al-Somood* 3 (31): 2.

Berejikian, Jeffrey D. 2002. "Model Building with Prospect Theory: A Cognitive Approach to International Relations." *Political Psychology* 23 (4): 759–786.

Bergen, Peter. 2001. *Holy War, Inc.: Inside the Secret World of Osama bin Laden*. New York: Simon & Schuster.

——. 2006. *The Osama Bin Laden I Know: An Oral History of Al Qaeda's Leader*. New York: Simon & Schuster.

Bergen, Peter, and Katherine Tiedemann. 2013. *Talibanistan: Negotiating the Borders Between Terror, Politics, and Religion*. New York: Oxford University Press.

Berko, Anat, and Edna Erez. 2005. ""Ordinary People" and "Death Work": Palestinian Suicide Bombers as Victimizers and Victims." *Violence and Victims* 20 (6): 603–623.

Bernbeck, Reinhard. 2010. "Heritage Politics: Learning from Mullah Omar?" In *Controlling the Past, Owning the Future: The Political Uses of Archaeology in the Middle East*, edited by Ran Boytner, Lynn Swartz Dodd, and Bradley J. Parker, 27–54. Tucson: University of Arizona Press.

Bhabha, Homi. 1994. *The Location of Culture*. New York: Routledge.

Bhui, Kamaldeep, and Yasmin Ibrahim. 2013. "Marketing the 'Radical': Symbolic Communication and Persuasive Technologies in Jihadist Websites." *Transcultural Psychiatry* 50 (2): 216–234.

Bibeau, Gilles. 1997. "Cultural Psychiatry in a Creolizing World: Questions for a New Research Agenda." *Transcultural Psychiatry* 34 (1): 9–41.

Bigg, Matthew. 2006. "U.S. 'Joe Blow' Keeps Track of Iraq War Dead." Reuters. December 27.

Bijleveld, Sophia Milosevic. 2008. "Afghanistan: Re-imagining the Nation Through the Museum—the Jihad Museum in Heart." *Studies in Ethnicity and Nationalism* 6 (2): 105–124.

Bockstette, Carsten. 2009. "Taliban and Jihadist Terrorist Use of Strategic Communication." *Connections* 3 (8): 1–24.

Bodansky, Yossef. 1999. *Bin Laden: The Man Who Declared War on America.* Roseville, Calif.: Prima Publishing.

Bonner, Michael. 2006. *Jihad in Islamic History: Doctrines and Practices.* Stanford: Stanford University Press.

Bose, Sugata, and Ayesha Jalal. 2004. *Modern South Asia: History, Culture, Political Economy.* New York: Routledge.

Boulding, Kenneth. 1959. "National Images and International Systems." *Journal of Conflict Resolution* 3 (2): 120–131.

Bowater, Margaret. 2012. "Dreams and Politics: How Dreams May Influence Political Decisions." *Psychotherapy and Politics International* 10 (1): 45–54.

Braithwaite, Rodric. 2011. *Afgantsy: The Russians in Afghanistan, 1979–89.* London: Profile Books.

Brown, Rupert. 2000. "Social Identity Theory: Past Achievements, Current Problems and Future Challenges." *European Journal of Social Psychology* 30 (6): 745–778.

Brown, Vahid. 2010. "The Façade of Allegiance: Bin Ladin's Dubious Pledge to Mullah Omar." *CTC Sentinel* 3 (1): 1–6.

Bunt, Gary R. 2000. *Virtually Islamic: Computer-Mediated Communication and Cyber Islamic Environments.* Cardiff: University of Wales Press.

——. 2009. *iMuslims: Rewiring the House of Islam.* Chapel Hill: University of North Carolina Press.

Burke, Jason. 2003. *Al-Qaeda: The True Story of Radical Islam.* New York: I.B. Taurus.

Canfield, Robert L. 2011. "Introduction: A Region of Strategic Importance." In *Ethnicity, Authority, and Power in Central Asia: New Games Great and Small,* edited by Robert L. Canfield and Gabriele Rasuly-Paleczek, 1–16. New York: Routledge.

Caron, James. 2012. "Taliban, Real and Imagined." In *Under the Drones: Modern Lives in the Afghanistan-Pakistan Borderlands*, edited by Shahzad Bashir and Robert D. Crews, 60–82. Cambridge: Harvard University Press.

Carpenter-Song, Elizabeth, Megan Nordquest Schwallie, and Jeffrey Longhofer. 2007. "Cultural Competence Reexamined: Critique and Directions for the Future." *Psychiatric Services* 58 (10): 1362–1365.

Centlivres, Pierre, and Micheline Centlivres-Demont. 2000. "State, National Awareness and Levels of Identity in Afghanistan from Monarchy to Islamic State." *Central Asian Survey* 19 (3/4): 419–428.

"Chand Chubtē Savālāt Kē Itmīnān Bakhsh Javābāt" [Comforting Answers to Piercing Questions] (Urdu). 2014. *Sharī'at* 3 (23): 16–21.

Chen, Hsinchun. 2012. *Exploring and Data Mining the Dark Side of the Web.* Heidelberg: Springer-Verlag GmbH.

Chen, Hsinchun, Wingyan Chung, Jialun Qin, Edna Reid, Marc Sageman, and Gabriel Weimann. 2008. "Uncovering the Dark Web: A Case Study of Jihad on the Web." *Journal of the American Society for Information Science and Technology* 59 (8): 1347–1359.

Cheung, Fanny M., Fons J. R. van de Vijver, and Frederick T. L. Leong. 2011. "Toward a New Approach to the Study of Personality in Culture." *American Psychologist* 66 (7): 593–603.

Chroust, Peter. 2000. "Neo-Nazis and Taliban On-line: Anti-Modern Political Movements and Modern Media." *Democratization* 7 (1): 102–118.

Church, A. Timothy. 2000. "Culture and Personality: Toward an Integrated Cultural Trait Psychology." *Journal of Personality* 68 (4): 651–703.

——. 2001. "Personality Measurement in Cross-Cultural Perspective." *Journal of Personality* 69 (6): 979–1006.

——. 2008. "Current Controversies in the Study of Personality Across Cultures." *Social and Personality Psychology Compass* 2 (5): 1930–1951.

Ciovacco, Carl J. 2009. "The Contours of Al Qaeda's Media Strategy." *Studies in Conflict and Terrorism* 32 (10): 853–875.

CNN.com. 2001. "UAE Cuts Ties with Taliban." September 22. Available at: http://edition.cnn.com/2001/WORLD/asiapcf/central/09/22/ret.afghan.taliban/.

Cohen, Stephen. 2004. *The Idea of Pakistan*. Washington, D.C.: Brookings Institution.

Cohler, Bertram J. 1992. "Intent and Meaning in Psychoanalysis and Cultural Study." In *New Directions in Psychological Anthropology*, edited by Theodore

Schwartz, Geoffrey M. White, and Catherine A. Lutz, 269–293. Cambridge: Cambridge University Press.

Coll, Steve. 2004. *Ghost Wars: The Secret History of the CIA, Afghanistan, and bin Laden, from the Soviet Invasion to September 10, 2001*. New York: Penguin Books.

——. 2011. "Looking for Mullah Omar." *New Yorker*, January 23, pp. 45–55.

Comaroff, John L., and Jean Comaroff. 2009. "Reflections on the Anthropology of Law, Governance and Sovereignty." In *Rules of Law and Laws of Ruling: On the Governance of Law*, edited by Franz von Benda-Beckmann, Keebet von Benda-Beckmann, and Julia Eckert, 31–60. Surrey: Ashgate.

Constitution (Second Amendment) Act, 1974. 1974. "The Constitution of Pakistan." Available at: http://www.pakistani.org/pakistan/constitution/amendments/2amendment.html, accessed August 12, 2014.

Cook, Susan E. 2004. "New Technologies and Language Change: Toward an Anthropology of Linguistic Frontiers." *Annual Review of Anthropology* 33:103–115.

Cottam, Martha L. 1985. "The Impact of Psychological Images on International Bargaining: The Case of Mexican Natural Gas." *Political Psychology* 6 (3): 413–440.

——. 1992. "The Carter Administration's Policy Toward Nicaragua: Images, Goals, and Tactics." *Political Science Quarterly* 107 (1): 123–146.

Cottam, Richard W. 1977. *Foreign Policy Motivation: A General Theory and a Case Study*. Pittsburgh: University of Pittsburgh Press.

Crapanzano, Vincent. 1992. "Some Thoughts on Hermeneutics and Psychoanalytic Anthropology." In *New Directions in Psychological Anthropology*, edited by Theodore Schwartz, Geoffrey M. White, and Catherine A. Lutz, 294–308. Cambridge: Cambridge University Press.

Crenshaw, Martha. 2000. "The Psychology of Terrorism: An Agenda for the 21st Century." *Political Psychology* 21 (2): 405–420.

——. 2004. "The Psychology of Political Terrorism." In *Political Psychology*, edited by John T. Jost and Jim Sidanius, 546–573. New York: Taylor & Francis Books.

Creswell, John W. 2014. *Research Design: Qualitative, Quantitative, and Mixed Methods Approaches*. Thousand Oaks: Sage.

Creswell, John W., and Dana L. Miller. 2000. "Determining Validity in Qualitative Inquiry." *Theory into Practice* 39 (3): 124–130.

Crone, Chase F. 2003. *Islamic Historiography*. Cambridge: Cambridge University Press.

Crone, Patricia, and Martin Hinds. 1996. *God's Caliph: Religious Authority in the First Centuries of Islam*. Cambridge: Cambridge University Press.

Cullison, Alan. 2004. "Inside Al-Qaeda's Hard Drive." *Atlantic*, September 1. Available at: http://www.theatlantic.com/magazine/archive/2004/09/inside-al-qaeda-s-hard-drive/303428/.

Dabashi, Hamid. 2012. *The World of Persian Literary Humanism*. Cambridge: Harvard University Press.

Davis, Anthony. 2001. "How the Taliban Became a Military Force." In *Fundamentalism Reborn? Afghanistan and the Taliban*, edited by William Maley, 43–71. New York University Press.

Dawi, Akmal. 2014. "Despite Massive Taliban Death Toll No Drop in Insurgency." *Voice of America*, March 6.

Dawood, Brother. 2013. "My Story: Journey of Brother Dawood." *Azan: A Call to Jihad* 1 (2): 73–81.

De Castella, Krista, Craig McGarty, and Luke Musgrove. 2009. "Fear Appeals in Political Rhetoric About Terrorism: An Analysis of Speeches by Australian Prime Minister Howard." *Political Psychology* 30 (1): 1–26.

Dekleva, Kenneth B., and Jerrold M. Post. 1997. "Genocide in Bosnia: The Case of Dr. Radovan Karadzic." *Journal of the American Academy of Psychiatry and the Law* 41 (3): 485–496.

Der Spiegel. 2004. Spiegel Interview: "And Then Mullah Omar Screamed at Me." Available at: http://www.spiegel.de/international/spiegel/spiegel-interview-and-then-mullah-omar-screamed-at-me-a-289592.html, accessed December 16, 2013.

Deutsch, Morton, and Catarina Kinnvall. 2002. "What Is Political Psychology?" In *Political Psychology*, edited by Kristen Renwick Monroe, 15–42. Mahwah: Lawrence Erlbaum.

"Diversionary Wars: Pashtun Unrest and the Sources of the Pakistan-Afghan Confrontation." 2011. *Canadian Foreign Policy Journal* 17 (1): 38–49.

Donner, Fred McGraw. 2010. *Muhammad and the Believers: At the Origins of Islam*. Cambridge: Harvard University Press.

Druckman, Daniel. 1994. "Nationalism, Patriotism, and Group Loyalty: A Social Psychological Perspective." *Mershon International Studies Review* 38 (1): 43–68.

Edgar, Iain R. 2006. "The 'True Dream' in Contemporary Islamic/Jihadist Dreamwork: A Case Study of the Dreams of Taliban Leader Mullah Omar." *Contemporary South Asia* 15 (3): 263–272.

——. 2011. *The Dream in Islam: From Qur'anic Tradition to Jihadist Inspiration*. Brooklyn: Bergahn Books.

Edgar, Iain, and David Henig. 2010. "*Istikhara*: The Guidance and Practice of Islamic Dream Incubation Through Ethnographic Comparison." *History and Anthropology* 21 (3): 251–262.

"Editorial." 2013. *Azan: A Call to Jihad* 1 (1): 3.

"Editorial." 2013. *Azan: A Call to Jihad* 1 (4): 5.

"Editorial." 2014. *Azan: A Call to Jihad* 1 (5): 4.

"Editorial." 2014. *Haqīqat* 1 (1): 2.

Edwards, David B. 1986. "Charismatic Leadership and Political Process in Afghanistan." *Central Asian Survey* 5 (3–4): 273–299.

——. 1995. "Print Islam: Media and Religious Revolution in Afghanistan." *Anthropological Quarterly* 68 (3): 171–184.

——. 2002. *Before Taliban: Genealogies of the Afghan Jihad*. Berkeley: University of California Press.

Efferson, Charles, Rafael Lalive, and Ernst Fehr. 2008. "The Coevolution of Cultural Groups and Ingroup Favoritism." *Science* 321:1844–1849.

Elcheroth, Guy, Willem Doise, and Stephen Reicher. 2011. "On the Knowledge of Politics and the Politics of Knowledge: How a Social Representations Approach Helps Us Rethink the Subject of Political Psychology." *Political Psychology* 32 (5): 729–758.

Entezar, Ehsan M. 2008. *Afghanistan 101: Understanding Afghan Culture*. Bloomington, Ind.: Xlibris.

Ernst, Carl W. 2011a. *How to Read the Qur'an: A New Guide, with Select Translations*. Chapel Hill: University of North Carolina Press.

——. 2011b. "The Limits of Universalism in Islamic Thought: The Case of Indian Religions." *Muslim World* 101 (1): 1–19.

Euben, Roxanne. 1995. "When Worldviews Collide: Conflicting Assumptions About Human Behavior Held by Rational Actor Theory and Islamic Fundamentalism." *Political Psychology* 16 (1): 157–178.

Fabrega, Horacio, Jr. 1994. "Personality Disorders as Medical Entities: A Cultural Interpretation." *Journal of Personality Disorders* 8 (2): 149–167.

Fairclough, Norman. 1992. *Discourse and Social Change*. Cambridge: Polity Press.

——. 2001. "Critical Discourse Analysis." In *How to Analyze Talk in Institutional Settings*, edited by Alec McHoul and Mark Rapley, 25–38. New York: Continuum.

Faiz, Ahmed Tariq. 2010. "Ayā Jehān Bakhsh Az Gharb Ast Yā Gharb Bakhsh Az Jehān?" [Is the World Part of the West or Is the West Part of the World?] (Dari). *Srak* 5 (56): 24–25.

Faizer, Rizwi. 2014. "Expeditions and Battles." In *Encyclopedia of the Qu'rān*, edited by Jane Dammen McAuliffe. Available at: http://referenceworks.brillonline.com/entries/encyclopaedia-of-the-quran/expeditions-and-battles-EQCOM_00060, accessed August 11, 2014.

Farmer, Ben. 2011. "Taliban Deny Mullah Omar Had Heart Attack." *Telegraph*, January 19. Available at: http://www.telegraph.co.uk/news/worldnews/asia/pakistan/8269275/Taliban-deny-Mullah-Omar-had-heart-attack.html.

Farr, Robert M. 1998. "From Collective to Social Representations: *Aller et Retour*." *Culture and Psychology* 4 (3): 275–296.

Farrell, Graham, and John Thorne. 2005. "Where Have All the Flowers Gone? Evaluation of the Taliban Crackdown Against Opium Poppy Cultivation in Afghanistan." *International Journal of Drug Policy* 16 (2): 81–91.

Farrell, Theo, and Antonio Giustozzi. 2013. "The Taliban at War: Inside the Helmand Insurgency, 2004–2012." *International Affairs* 89 (4): 845–871.

Fassin, Didier. 2008. "Beyond Good and Evil: Questioning the Anthropological Discomfort with Morals." *Anthropological Theory* 8 (4): 333–344.

Fergusson, James. 2010. *Taliban: The Unknown Enemy*. Boston: Da Capo Press.

"Fī Intidhār 'Ghadhba 'Uthmānīya' Min Al-Raīs Al-Turkī" [Waiting for the "Ottoman Rage" from the Turkish President] (Arabic). 2009. *Al-Somood* 3 (42): 39.

Filkins, Dexter. 2001. "Rise and Fall; The Legacy of the Taliban Is a Sad and Broken Land." *New York Times*, December 31, p. 1.

——. 2010. "Pakistanis Tell of Motive in Taliban Leader's Arrest." *New York Times*, August 22.

Financial Commission. 2012. "The Islamic Emirate of Afghanistan Financial Commission." *Islamic Emirate of Afghanistan*. Available at: http://shahamat-english.com/english/index.php/paighamoona/28810, accessed January 7, 2014.

Finlayson, Alan. 1998. "Psychology, Psychoanalysis, and Theories of Nationalism." *Nations and Nationalism* 4 (2): 145–162.

Fiorillo, Andrea, Mario Luciano, Valeria Del Vecchio, Gaia Sampogna, Carla Obradors-Tarragó, and Mario Maj, on behalf of the ROAMER Consortium. "Priorities for Mental Health Research in Europe: A Survey Among National Stakeholders' Associations Within the ROAMER Project." *World Psychiatry* 12 (2): 165–170.

Foucault, Michel. 1970. *The Order of Things: An Archaeology of the Human Sciences*. New York: Vantage Books.

——. 1980. *Power/Knowledge: Selected Interviews and Other Writings, 1972–1977*. Edited by Colin Gordon. New York: Vintage Books.

——. 1990. *The History of Sexuality: Volume 1; An Introduction*. New York: Vantage Books.

——. 1991. "Politics and the Study of Discourse." In *The Foucault Effect: Studies in Governmentality*, edited by Graham Burchell, Colin Gordon, and Peter Miller, 53–72. Chicago: University of Chicago Press.

——. 1994. *The Order of Things: An Archaeology of the Human Sciences*. New York: Vintage Books.

Freud, Sigmund. 1913. *The Interpretation of Dreams*. 3rd ed. Translated by A. A. Brill. New York. Macmillan

Galander, Mahmoud M. 2002. "Communication in the Early Islamic Era: A Social and Historical Analysis." *Intellectual Discourse* 10 (1): 61–75.

Gamson, William A., and Hanna Herzog. 1999. "Living with Contradictions: The Taken-for-Granted in Israeli Political Discourse." *Political Psychology* 20 (2): 247–266.

Gargan, Edward A. 2001. "A Lavish Lifestyle—for an Afghan." *Baltimore Sun*, December 15. Available at: http://articles.baltimoresun.com/2001-12-15/news/0112150216_1_mullah-mohammed-omar-taliban-leader-mullah-lifestyle.

Geertz, Clifford. 2005. "Shifting Aims, Moving Targets: On the Anthropology of Religion." *Journal of the Royal Anthropological Institute* 11 (1): 1–15.

Ghani, Ashraf. 1978. "Islam and State-Building in a Tribal Society Afghanistan: 1880–1901." *Modern Asian Studies* 12 (2): 269–284.

Ghaznavi, Shihabuddin. 2008a. "Al-Furūq Al-Jauharīya Baina Al-Ihtilāl Al-Amrīkī Wa Al-Ihtilāl Al-Rūsī" [The Essential Differences Between the American Occupation and the Soviet Occupation] (Arabic). *Al-Somood* 3 (25): 19–22.

——. 2008b. "Al-Mukhaddirāt Hīya Al-Nau' Al-Ākhar Min Al-Asliha Al-Latī Tasktakhdimu-hā Amrīka Li-Dharb Al-Sha'ab Al-Afghānī" [Intoxicants Are the Latest Type of Weapons That America Uses to Strike the Afghan People] (Arabic). *Al-Somood* 3 (27): 44–46.

——. 2008c. "Hazīma Amrīka Fī Afghānistān Satakūn Asra' Min Hazīma Al-Ittihād Al-Sofiātī Al-Munhār: Shaikh Jalāluddīn Haqqānī" [The Defeat of America in Afghanistan Will Be Quicker Than the Defeat of the Collapsed Soviet Union: Sheikh Jalaluddin Haqqani] (Arabic). *Al-Somood* 3 (30): 9–15.

——. 2009. "Harkat Tālibān Nadharīya Islāmīya Dhātu Judhūr 'amīqa" [The Taliban Movement Is Ideologically Islamic with Deep Roots] (Arabic). *Al-Somood* 3 (33): 16–17.

Gillespie, Alex, Caroline S. Howarth, and Flora Cornish. 2012. "Four Problems for Researchers Using Social Categories." *Culture and Psychology* 18 (3): 391–402.

Giustozzi, Antonio. 2008. *Koran, Kalashnikov, and Laptop: The Neo-Taliban Insurgency in Afghanistan*. New York: Columbia University Press.

——. 2009. *Empires of Mud: Wars and Warlords in Afghanistan*. New York: Columbia University Press.

——. 2010. *The Taliban Beyond the Pashtuns*. Afghanistan Papers 5. Waterloo, Ont.: Centre for International Governance Innovation.

——. 2012. *Decoding the New Taliban: Insights from the Afghan Field*. New York: Oxford University Press.

Glenn, Edmund S., Robert H. Johnson, Paul R. Kimmel, and Bryant Wedge. 1970. "A Cognitive Interaction Model to Analyze Culture Conflict in International Relations." *Journal of Conflict Resolution* 14 (1): 35–48.

Gohari, M. J. 1999. *The Taliban: Ascent to Power*. New York: Oxford University Press.

Goldenberg, Suzanne. 1998. "Heart of Darkness." *Guardian*, October 13.

Goldstein, Joseph. 2015. "In ISIS, the Taliban Face an Insurgent Threat of Their Own." *New York Times*, June 4.

Gommans, Jos. J. L. 1999. *The Rise of the Indo-Afghan Empire, c. 1710–1780*. Delhi: Oxford University Press.

Good, Byron. 1996. "Culture and DSM-IV: Diagnosis, Knowledge and Power." *Culture, Medicine, and Psychiatry* 20 (2): 127–132.

Goodson, Larry P. 2012. *Afghanistan's Endless War: State Failure, Regional Politics, and the Rise of the Taliban*. Seattle: University of Washington Press.

Gopal, Anand. 2010. "Half of Afghanistan Taliban Leadership Arrested in Pakistan." *Christian Science Monitor*, February 24. Available at: http://www.csmonitor.com/World/Asia-South-Central/2010/0224/Half-of-Afghanistan-Taliban-leadership-arrested-in-Pakistan.

Gray, Debra, and Kevin Durrheim. 2013. "Collective Rights and Personal Freedoms: A Discursive Analysis of Participant Accounts of Authoritarianism." *Political Psychology* 34 (4): 631–648.

Green, Nile. 2003. "The Religious and Cultural Roles of Dreams and Visions in Islam." *Journal of the Royal Asiatic Society of Great Britain and Ireland* 13 (3): 287–313.

——. 2010. "The Dilemmas of the Pious Biographer: Missionary Islam and the Oceanic Hagiography." *Journal of Religious History* 34 (4): 383–397.

——. 2011. "The Trans-Border Traffic of Afghan Modernism: Afghanistan and the Indian 'Urdusphere.'" *Comparative Studies in Society and History* 53 (3): 479–508.

Guarnaccia, Peter J., and Orlando Rodriguez. 1996. "Concepts of Culture and Their Role in the Development of Culturally Competent Mental Health Services." *Hispanic Journal of Behavioral Sciences* 18 (4): 419–443.

Guest, Greg, Arwen Bunce, and Laura Johnson. 2006. "How Many Interviews Are Enough? An Experiment with Data Saturation and Variability." *Field Methods* 18 (1): 59–82.

Gumperz, John J., and Jenny Cook-Gumperz. 2008. "Studying Language, Culture, and Society: Sociolinguistics or Linguistic Anthropology." *Journal of Sociolinguistics* 12 (4): 532–545.

Gunaratna, Rohan. 2002. *Inside Al Qaeda: Global Network of Terror*. New York: Columbia University Press.

Gupta, Akhil, and James Ferguson. 1992. "Beyond "Culture": Space, Identity, and the Politics of Difference." *Cultural Anthropology* 7 (1): 6–23.

Gutman, Roy. 2010. "Afghanistan War: How Taliban Tactics Are Evolving." *Christian Science Monitor*, March 15.

Guzzini, Stefano. 2000. "A Reconstruction of Constructivism in International Relations." *European Journal of International Relations* 6 (2): 147–182.

Habib, Qari. 2014. "Wa Harb Al-Ghazāh Al-Kanadīūna" [And the Flight of the Canadian Expedition] (Arabic). *Al-Somood* 8 (96): 4–5.

Hakimzad, Abdul Rahman. 2010. "Che Gūne Farzandān-e Khudrā Az Mufāsad-e Ikhlāqī Nigahdārī Nemāyīm" [How Do We Stop the Moral Corruption of Our Children?] (Dari). *Srak* 5 (56): 49–51.

Hāmid, Mustafa. 2012a. "Amrīkā Tālibān Kē Narghē Men" [America in the Press of the Taliban] (Urdu). *Sharī'at* 1 (6): 27–34.

——. 2012b. "Muslimī Mīanmār Yatajarra'ūna Kasa Al-Dīmuqrātīya" [The Muslims of Myanmar Drink from the Cup of Democracy] (Arabic). *Al-Somood* 7 (76): 6–10.

Hammack, Phillip L. 2008. "Narrative and the Cultural Psychology of Identity." *Personality and Social Psychology Review* 12 (3): 222–247.

——. 2010. "The Cultural Psychology of Palestinian Youth: A Narrative Approach." *Culture and Psychology* 16 (4): 507–537.

Hammack, Phillip L. and Andrew Pilecki. 2012. "Narrative as a Root Metaphor for Political Psychology." *Political Psychology* 33 (1): 75–103.

Hammad, Hanif. 2012. "Pamparz Wāle Amrīkī Faujī" [Pampered American Soldiers] (Urdu). *Sharī'at* 1 (1): 16–17.

Hammadi, Badruddin. 2007. "Al-Qadhīyya Al-Filistīnīya Alān Fī Ahsan Marahil-hā" [The Palestinian Issue Now Is in Its Best Phase] (Arabic). *Al-Somood* 2 (13): 30.

Haqqani, Husain. 2008. "Insecurity Along the Durand Line." In *Afghanistan: Transition Under Threat*, edited by Geoffrey Hayes and Mark Sedra, 219–237. Waterloo, Ont.: Wilfrid Laurier University Press.

Haroon, Sana. 2007. *Frontier of Faith: Islam in the Indo-Afghan Borderland*. New York: Columbia University Press.

——. 2008. "The Rise of Deobandi Islam in the North-West Frontier Province and Its Implications in Colonial India and Pakistan 1914–1996." *Journal of the Royal Asiatic Society* 18 (1): 47–70.

Harpviken, Kristian Berg. 2012. "The Transnationalization of the Taliban." *International Area Studies Review* 15 (3): 203–229.

Heine, Steven J., and Emma E. Buchtel. 2009. "Personality: The Universal and the Culturally Specific." *Annual Review of Psychology* 60: 369–394.

Hermann, Margaret G. 1980. "Assessing the Personalities of Soviet Politburo Members." *Personality and Social Psychology Bulletin* 6 (3): 332–352.

——. 2003. "Assessing Leadership Style: Trait Analysis." In *The Psychological Assessment of Political Leaders: With Profiles of Saddam Hussein and Bill Clinton*, edited by Jerrold Post, 178–212. Ann Arbor: University of Michigan Press.

Hermann, Richard K. 1985. "Analyzing Soviet Images of the United States: A Psychological Theory and Empirical Study." *Journal of Conflict Resolution* 29 (4): 665–697.

Hermann, Richard K., and Michael P. Fischerkeller. 1995. "Beyond the Enemy Image and Spiral Model: Cognitive-Strategic Research After the Cold War." *International Organization* 49 (3): 415–450.

Hermann, Richard K., Pierangelo Isernia, and Paolo Segatti. 2009. "Attachment to the Nation and International Relations: Dimensions of Identity and Their Relationship to War and Peace." *Political Psychology* 30 (5): 721–754.

Hikmat, Abdul Rauf. 2012a. "Burmā Kē Sānha-e Fāji Kē Mutallaq Imārat-e Islāmī Kā Zimedārana Aur Tanbīhī Peghām Aur Dunyā Par Is Kē Asarāt" [The Responsible and Warning Message of the Islamic Emirate Regarding the Emergent Tragedy of Burma and Its Effects on the World] (Urdu). *Sharī'at* 1 (6): 9–10.

——. 2012b. "Shahīd Mulla Saifur Rahmān Mansūr Rahmatullah Alehi Kī Shahādat Kā Dasvān Sāl" [The Tenth Anniversary of the Martyrdom of the Martyr Mullah Saifur Rahman Mansur, May God Have Mercy Upon Him] (Urdu). *Sharī'at* 1 (3): 16–21.

——. 2013a. "Hāfiz Badruddīn Haqqānī Shahīd Kē Hayāt Aur Kānāmon Par Ek Nazar" [A Look at the Life and Accomplishments of the Martyr Hafiz Badruddin Haqqani] (Urdu). *Sharī'at* 2 (22): 8–10.

——. 2013b. "Maulawī Abdul Hanān Jihādwāl Shahīd Kī Hayāt Aur Kārnāmon Par Ek Nazar" [A Look at the Life and Accomplishments of the Martyr Maulawi Abdul Hanan Jihadwal] (Urdu). *Sharī'at* 2 (15): 4–7.

Hobsbawm, Eric. 1996. "Language, Culture, and National Identity." *Social Research* 63 (4): 1065–1080.

Hodgson, Marshall G. S. 1977. *The Venture of Islam: Conscience and History in a World Civilization. Volume 1: The Classical Age of Islam*. Chicago: University of Chicago Press.

Hofstede, Geert, and Robert R. McCrae. 2004. "Personality and Culture Revisited: Linking Traits and Dimensions of Culture." *Cross-Cultural Research* 38 (1): 52–88.

Hogg, Michael A., and Scott A. Reid. 2006. "Social Identity, Self-Categorization, and the Communication of Group Norms." *Communication Theory* 16 (1): 7–30.

Hogg, Michael A., Deborah J. Terry, and Katherine M. White. 1995. "A Tale of Two Theories: A Critical Comparison of Identity Theory with Social Identity Theory." *Social Psychological Quarterly* 58 (4): 255–269.

Holzman, John C. 1995. "Afghanistan: Russian Embassy Official Claims Iran Interfering More Than Pakistan." National Security Archives. January 13. Available at: www2.gwu.edu/~nsarchiv/NSAEBB/NSAEBB227/6.pdf.

Horgan, John. 2004. *The Psychology of Terrorism*. New York: Routledge.

——. 2008. "Interviewing Terrorists." In *Terrorism Informatics: Knowledge Management and Data Mining for Homeland Security*, edited by Hsinchun Chen, Edna Reid, Joshua Sinai, Andrew Silke, and Boaz Ganor, 73–99. New York: Springer-Verlag.

Houghton, David Patrick. 1998. "Historical Analogies and the Cognitive Dimension of Domestic Policymaking." *Political Psychology* 19 (2): 279–303.

Howarth, Caroline. 2002. "Identity in Whose Eyes? The Role of Representations in Identity Construction." *Journal for the Theory of Social Behaviour* 32 (2): 145–162.

Huddy, Leonie. 2001. "From Social to Political Identity: A Critical Examination of Social Identity Theory." *Political Psychology* 22 (1): 127–156.

Huggler, Justin. 2001. "Inside Mullah Omar's Palace of Kitsch." *New Zealand Herald*, December 13. Available at: http://www.nzherald.co.nz/world/news/article.cfm?c_id=2&objectid=233743.

Humphreys, R. Stephen. 1991. *Islamic History: A Framework for Inquiry*. Princeton, N.J.: Princeton University Press.

Hussain, Jaffer. 2013. "The American Gorbachev." *Azan: A Call to Jihad* 1 (2): 33–36.

——. 2014. "Counter-Drone Strategy: Join the Caravan." *Azan: A Call to Jihad* 1 (5): 8–10.

Hussain, Rizwan. 2005. *Pakistan and the Emergence of Islamic Militancy in Afghanistan*. Burlington: Ashgate.

Hyman, Anthony. 2002. "Nationalism in Afghanistan." *International Journal of Middle East Studies* 34 (2): 299–315.

iCasualties. 2006. "Fatalities by Year and Month." Available at: icasualties.org/OEF/ByMonth.aspx, accessed December 30, 2014.

"Ihsāīyāt Al-Jihād" [The Statistics of Jihad] (Arabic). 2006. *Al-Somood* 1 (1): 32.

Imran, Ustadh Abu. 2014. "My Story." *Azan: A Call to Jihad* 1 (6): 34–38.

Inskeep, Steve. 2002. "The Cloak of the Prophet: Religious Artifact at the Heart of Former Taliban Stronghold." *National Public Radio*, January 10.

"Interview with Mullah Omar—Transcript." November 15, 2001. BBC News. Available at: http://news.bbc.co.uk/2/hi/south_asia/1657368.stm.

Islamic Emirate of Afghanistan. 1998a. "The Movement of Taliban." www.taliban.com. Available at: https://web.archive.org/web/19980623170905/http://www.taliban.com/talibmov.htm, accessed February 6, 2015.

——. 1998b. "Offer Your Qurbani in Kabul." www.taliban.com. Available at: https://web.archive.org/web/19980623165745/http://www.taliban.com/support.htm, accessed February 6, 2015.

——. 1998c. "Sheikh-ul-Hadith Hadhrat Maulana Muhammad Moosa Rohani Bazi's Declaration of Faith in the Taliban." www.taliban.com. Available at: https://web.archive.org/web/19980623171156/http://www.taliban.com/faith.htm, accessed February 6, 2015.

——. 2006a. "Bayān Amīr-ul-Muminīn Bi-Munāsaba Al-'Udwān Al-Israīlī 'Ala Lubnān" [Statement of the Commander of the Faithful Regarding the Israeli Aggression Against Lebanon] (Arabic). *Al-Somood* 1 (2): 2.

——. 2006b. "Kalima Amīr-ul-Muminīn Mullah Muhammad 'Umar Fī Rithā Shahīd Al-Umma Al-Islāmīya Al-Mujāhid Abū Musa'b Al-Zarqāwī" [The Speech of the Commander of the Faithful Mullah Muhammad Omar Regarding the Demise of the Martyr of the Islamic Community, the Mujahid Abu Musab Al-Zarqawi] (Arabic). *Al-Somood* 1 (1): 4.

——.2012a. "Farmān-e-Amīr-ul-Muminīn" [Order of the Commander of the Faithful] (Urdu). *Sharī'at: Monthly Islamic Magazine* 1 (1): 49.

——. 2012b. "Farmān-e-Amīr-ul-Muminīn" [Order of the Commander of the Faithful] (Urdu). *Sharī'at: Monthly Islamic Magazine* 1 (2): 51.

——.2012c. "Farmān-e-Amīr-ul-Muminīn" [Order of the Commander of the Faithful] (Urdu). *Sharī'at: Monthly Islamic Magazine* 1 (6): 56.

——. 2012d. "Farmān-e-Amīr-ul-Muminīn" [Order of the Commander of the Faithful] (Urdu). *Sharī'at: Monthly Islamic Magazine* 1 (7): 51.

——. 2012e. "Farmān-e-Amīr-ul-Muminīn" [Order of the Commander of the Faithful] (Urdu). *Sharī'at: Monthly Islamic Magazine* 1 (8): 51.

——. 2012f. "Farmān-e-Amīr-ul-Muminīn" [Order of the Commander of the Faithful] (Urdu). *Sharī'at: Monthly Islamic Magazine* 1 (9): 51.

——. 2013a. "All-Inclusive Interview by *Morchal* Magazine with the Deputy Head of the Cultural Commission of the Islamic Emirate." Available at: http://shahamat-english.com/english/index.php/interviwe/32677-all-inclusive-interview-by-'morchal'-magazine-with-the-deputy-head-of-the-cultural-commission-of-the-islamic-emira, accessed January 31, 2014.

——. 2013b. "Farmān-e-Amīr-ul-Muminīn" [Order of the Commander of the Faithful] (Urdu). *Sharī'at: Monthly Islamic Magazine* 1 (11): 51.

——. 2013c. "Farmān-e-Amīr-ul-Muminīn" [Order of the Commander of the Faithful] (Urdu). *Sharī'at: Monthly Islamic Magazine* 2 (17): 51.

——. 2013d. "Farmān-e-Amīr-ul-Muminīn" [Order of the Commander of the Faithful] (Urdu). *Sharī'at: Monthly Islamic Magazine* 2 (19): 51.

——. 2013e. "Farmān-e-Amīr-ul-Muminīn" [Order of the Commander of the Faithful] (Urdu). *Sharī'at: Monthly Islamic Magazine* 2 (21): 51.

——. 2013f. "Message of Felicitation of Amir-ul-Momineen (May Allah Protect Him) on the Occasion of Eid-ul-Fitr." August 6. Available at: http://www.shahamat-english.com/english/index.php/paighamoona/35234-message-of-felicitation-of-amir-ul-momineen-may-allah-protect-him-on-the-occasion-of-eid-ul-fitr, accessed December 23, 2013.

——. 2014a. "Al-Sādiqīn" [The Sincere] (Dari). Available at: http://shahamat-movie.com/index.php?option=com_webplayer&view=video&wid=115, accessed September 6, 2014.

——. 2014b. "Alemarah 3." Available at: http://shahamat-movie.com/index.php?option=com_webplayer&view=video&wid=37, accessed September 5, 2014.

——. 2014c. "An Interview with the Head of 'Financial Commission' of the Islamic Emirate." Available at: http://shahamat-english.com/index.php/interviwe/45804, accessed July 5, 2014.

——. 2014d. "Jihād" (Urdu). Available at: http://shahamat-movie.com/index.php?option=com_webplayer&view=video&wid=56, accessed September 7, 2014.

——. 2014e. "Katāib Badr 3" [The Badr Brigades 3] (Arabic). Available at: http://shahamat-movie.com/index.php/ویدیف-یبرع, accessed September 5, 2014.

——. 2015a. "Statement by the Leadership Council and Family of the Deceased Regarding the Passing Away of Amir ul Mumineen Mulla Muhammad Omar Mujahid: May Allah Have Mercy on Him." Available at: http://shahamat-english.com/statement-by-the-leadership-council-and-family-of-the-deceased-regarding-the-passing-away-of-amir-ul-mumineen-mulla-muhammad-omar-mujahid-may-allah-have-mercy-on-him/, accessed September 11, 2015.

——. 2015b. "Introduction of the Newly Appointed Leader of Islamic Emirate, Mullah Akhtar Mohammad (Mansur), May Allah Safeguard Hi[m]." Available at http://shahamat-english.com/introduction-of-the-newly-appointed-leader-of-islamic-emirate-mullah-akhtar-mohammad-mansur-may-allah-safeguard-hi/, accessed September 11, 2015.

Jalal, Qazi Abdul. 2013. "Maulawi Abdul Qudus." *Srak* 8 (76): 85–86.

Janbaz, Muhammad Farhad. 2013a. "Rūs Pareshān Ho Gayā Hai" [Russia Has Become Anxious] (Urdu). *Sharī'at* 2 (20): 27–29.

——. 2013b. "Shām Kā Khūn Āshām" [Syria's Blood Thirst] (Urdu). *Sharī'at* 2 (15): 41–43.

Jervis, Robert. 1976. *Perception and Misperception in International Politics*. Princeton, N.J.: Princeton University Press.

——. 2002. "Signaling and Perception: Drawing Inferences and Projecting Images." In *Political Psychology*, edited by Kristen Renwick Monroe, 293–312. Mahwah: Lawrence Erlbaum.

Johnson, Scott, and Evan Thomas. 2002. "Mulla Omar off the Record." *Newsweek*, January 21, p. 26.

Johnson, Thomas H. 2013. "Taliban Adaptations and Innovations." *Small Wars and Insurgencies* 24 (1): 3–27.

Johnson, Thomas H., and Matthew C. DuPee. 2012. "Analysing the New Taliban Code of Conduct (Layeha): An Assessment of Changing perspectives and Strategies of the Afghan Taliban." *Central Asian Survey* 31 (1): 77–91.

Johnson, Thomas H., and Ahmad Waheed. 2011. "Analyzing Taliban *Taranas* (Chants): An Effective Afghan Propaganda Artifact." *Small Wars and Insurgencies* 22 (1): 3–31.

Jost, John T., and Jim Sidanius. 2004. "Political Psychology: An Introduction." In *Political Psychology*, edited by John T. Jost and Jim Sidanius, 1–21. New York: Taylor & Francis.

"Kābul Kī Mujawaza Ulamā Kānfarans Aur Ulamā-e Haqq Se Muadabāne Apīl" [The Authorized Conference of Religious Scholars of Kabul and a Respectful Appeal to True Religious Scholars] (Urdu). 2013. *Sharī'at* 1 (10): 2.

"Kābul Men Munaqad Honē Walī Ulamā Kānfarans Kē Mutalaq Imārat-e Islāmīya Kā Afghānistān, Saudī Arab, Pākistān, Aur Ulamā-e Deoband Kē Nām Khasūsī Khat" [A Special Letter to the Religious Scholars of Afghanistan, Saudi Arabia, Pakistan, and Deoband from the Islamic Emirate Related to the Conference of Religious Scholars Being Organized in Kabul] (Urdu). 2013. *Sharī'at* 1 (10): 11–12.

Kakar, M. Hassan. 1995. *Afghanistan: The Soviet Invasion and the Afghan Response, 1979–1982*. Berkeley: University of California Press.

"Karzāī Aur Obāmā Kē Mā Bēn Istratijīk Muāhidē Par Dastakhat Par Imārat-e Islāmīya Kā Radd-e Amal" [The Reaction of the Islamic Emirate upon the Signing of the Strategic Agreement Between Karzai and Obama] (Urdu). 2012. *Sharī'at* 1 (3): 28.

Kavkaz Center. 2014. "About Kavkaz Center (The Caucasus Center)." Available at: http://www.kavkazcenter.com/eng/about/, accessed August 2, 2014.

Kennedy, Hugh. 2004. *The Prophet and the Age of the Caliphates, Second Edition*. Harlow: Pearson Education.

Khan, Zaheer. 2012. "Amrīka Ānkhon Men Dhūl Jhūnknē Kī Bajāē Maslē Kē Asal Hall Kī Jānib Qadam Uthāē" [America Should Take Steps Toward the Real Solution of the Problem Rather Than Blowing Dust in Eyes] (Urdu). *Sharī'at* 1 (1): 40–41.

Khel, Painda. 2012. "Nigāh-e Buland-Sukhan Dilnawāz Jān Par Sōz: Mulla Muhammad 'Umar Mujāhid" [A Burn upon the Tender Soul on Looking at High Speech: Mullah Muhammad Umar Mujahid] (Urdu). *Sharī'at* 1 (2): 22–23.

Khorasani, Hafiz Muhammad. 2013. "Tajlīl-e Jashan-e Nawrūz" [Celebrating the Festival of Nawrūz] (Dari). *Srak* 8 (73): 39–41.

Kinnvall, Catarina, and Paul Nesbitt-Larking. 2011. *The Political Psychology of Globalization: Muslims in the West*. Cambridge: Cambridge University Press.

Kirmayer, Laurence. 2007. "Psychotherapy and the Cultural Concept of the Person." *Transcultural Psychiatry* 44 (2): 232–257.

Kirmayer, Laurence J. 2006. "Beyond the 'New Cross-cultural Psychiatry': Cultural Biology, Discursive Psychology and the Ironies of Globalization." *Transcultural Psychiatry* 43 (1): 126–144.

Kirmayer, Laurence J., and Harry Minas. 2000. "The Future of Cultural Psychiatry: An International Perspective." *Canadian Journal of Psychiatry* 45 (5): 438–446.

Kirmayer, Laurence J., Eugene Raikhel, and Sadeq Rahimi. 2013. "Cultures of the Internet: Identity, Community, and Mental Health." *Transcultural Psychiatry* 50 (2): 165–191.

Kirmayer, Laurence J., Cécile Rousseau, and Jaswant Guzder. 2014. "Introduction: The Place of Culture in Mental Health Services." In *Cultural Consultation: Encountering the Other in Mental Health Care*, edited by Laurence J. Kirmayer, Jaswant Guzder, and Cécile Rousseau, 1–20. New York: Springer.

Kleiner, Jürgen. 2000. "The Taliban and Islam." *Diplomacy and Statecraft* 11 (1): 19–32.

Knysh, Alexander. 2012. "Islam and Arabic as the Rhetoric of Insurgency: The Case of the Caucasus Emirate." *Studies in Conflict and Terrorism* 35 (4): 315–337.

"Konfarans-e Ulemā Qabl Az Iniqād Nākām Gardīd" [The Conference of Religious Scholars Became Unsuccessful Before Convening] (Dari). 2013. *Srak* 8 (73): 57–58.

Kort, Alexis. 2005. "Dar al-Cyber Islam: Women, Domestic Violence, and the Islamic Reformation on the World Wide Web." *Journal of Muslim Minority Affairs* 25 (3): 363–383.

Kruglanski, Arie W., Xiaoyan Chen, Mark Dechesne, Shira Fishman, and Edward Orehek. 2009a. "Fully Committed: Suicide Bombers' Motivation and the Quest for Personal Significance." *Political Psychology* 30 (3): 331–357.

——. 2009b. "Yes, No, and Maybe in the World of Terrorism Research: Reflections on the Commentaries." *Political Psychology* 30 (3): 401–417.

Kushev, Vladimir. 1997. "Area Lexical Contacts of the Afghan (Pashto) Language (Based on the Texts of the XVI–XVIII Centuries)." *Iran and the Caucasus* 1 (1): 159–166.

"Kyā Yeh Dehshat Gardī Nahīn?" [Is This Not Terrorism?] (Urdu). 2012. *Sharī'at* 1 (1): 31–32.

Lacan, Jacques. 2006. *Ecrits*. New York: W. W. Norton.

Laghari, Nazeer, and Mufti Jameel Khan. 1998. "Interview with the Ameerul M'umineen." www.taliban.com. Available at: https://web.archive.org/web/19980623170926/http://www.taliban.com/intview1.htm, accessed December 20, 2013.

Lamb, Christina. 2001. "The Taliban Chief Is Mad, Says His Doctor." *Sunday Telegraph* (London), October 7, p. 4.

Lamoreaux, John C. 2002. *The Early Muslim Tradition of Dream Interpretation*. Albany: State University of New York Press.

Lanham, Trevor. 2011. "Mullah Muhammed Omar: A Psychobiographical Profile." *Culture and Conflict Review* 5 (3). Available at: http://nps.edu/programs/ccs/webjournal/Article.aspx?ArticleID=87, accessed August 10, 2013.

Lapidus, Ira M. 1996. "State and Religion in Islamic Societies." *Past and Present* 151: 3–27.

Lappin, Yaakov. 2011. *Virutal Caliphate: Exposing the Islamist State on the Internet*. Washington, D.C.: Potomac Books, Inc.

Larkin, Margaret. 2008. *Al-Mutanabbi: The Poet of Sultans and Sufis*. London: Oneworld Publications.

Lawrence, Bruce. 2005. *Messages to the World: The Statements of Osama bin Laden*. New York: Verso

Lehman, Darrin R., Chi-yue Chiu, and Mark Schaller. 2004. "Psychology and Culture." *Annual Review of Psychology* 55:689–714.

LeVine, Robert A. 2001. "Culture and Personality Studies, 1918–1960: Myth and History." *Journal of Personality* 69 (6): 803–818.

Liebl, Vern. 2007. "Pushtuns, Tribalism, Leadership, Islam and Taliban: A Short View." *Small Wars and Insurgencies* 18 (3): 492–510.

Littlewood, Roland. 1992. "DSM-IV and Culture: Is the Classification Internationally Valid?" *Psychiatric Bulletin* 16:257–261.

——. 1993. "Ideology, Camouflage, or Contingency? Racism in British Psychiatry." *Transcultural Psychiatry* 30 (3): 243–290.

——. 1996. "Psychiatry's Culture." *International Journal of Social Psychiatry* 42 (4): 245–268.

Liu, James H., Chris G. Sibley, and Li-Li Huang. 2013. "History Matters: Effects of Culture Specific Symbols on Political Attitudes and Intergroup Relations." *Political Psychology* 35 (1): 57–79.

LoCicero, Alice, and Samuel J. Sinclair. 2008. "Terrorism and Terrorist Leaders: Insights from Developmental and Ecological Psychology." *Studies in Conflict and Terrorism* 31 (3): 227–250.

Loidolt, Bryce. 2011. "Managing the Global and Local: The Dual Agendas of Al Qaeda in the Arabian Peninsula." *Studies in Conflict and Terrorism* 34 (2): 102–123.

"Ma'ādin-e Afghānistān Dar Changāl-e Ghāratgarān" [The Minerals of Afghanistan in the Hands of Pillagers] (Dari). 2013. *Srak* 8 (78): 14.

Maass, Anne, Federica Politi, Minoru Karasawa, and Sayaka Suga. 2006. "Do Verbs and Adjectives Play Different Roles in Different Cultures? A Cross-Linguistic Analysis of Person Representation." *Journal of Personality and Social Psychology* 90 (5): 734–750.

Magnus, Ralph H., and Eden Naby. 2002. *Afghanistan: Mullah, Marx, and Mujahid*. Boulder, Colo.: Westview Press.

Maiwandi, Akram. 2007a. "Ma'a Al-Shuhadā" [With the Martyrs] (Arabic). *Al-Somood* 1 (9): 18–20.

——. 2007b. "Shuhadā-Nā Al-Abtāl" [Our Martyrs, The Heroes] (Arabic). *Al-Somood* 1 (11): 18–20.

——. 2010. "Wājib Al-'Alam Al-Islāmī Tijāh Afghānistān" [The Duty of the Islamic World Toward Afghanistan] (Arabic). *Al-Somood* 5 (53): 3–5.

Makdisi G. "Ibn Kudāma al-Makdīsī." *Encyclopedia of Islam,* 2nd ed. Edited by P. Bearman, Th. Blanquis, C. E. Bosworth, E. van Donzel, and W. P. Heinrichs. Available at: http://referenceworks.brillonline.com/entries/encyclopaedia-of-islam-2/ibn-k-uda-ma-al-mak-di-si-SIM_3262, accessed February 11, 2015.

Malik, Hafeez. 1999. "The Taliban's Islamic Emirate of Afghanistan: Its Impact on Eurasia." *Brown Journal of World Affairs* 6 (1): 135–146.

Malkasian, Carter. 2013. *War Comes to Garmser: Thirty Years of Conflict on the Afghan Frontier*. New York: Oxford University Press.

Markus, Hazel Rose, and Shinobu Kitayama. 1998. "The Cultural Psychology of Personality." *Journal of Cross-Cultural Psychology* 29 (1): 63–87.

Marsden, Peter. 1998. *The Taliban: War, Religion, and the New World Order in Afghanistan*. New York: Zed Books.

Matinuddin, Kamal. 1999. *The Taliban Phenomenon Afghanistan 1994–1997: With an Afterword Covering Major Events Since 1997*. Karachi: Oxford University Press.

McCauley, Clark. 2001. "The Psychology of Group Identification and the Power of Ethnic Nationalism." In *Ethnopolitical Warfare: Causes, Consequences, and Possible Solutions*, edited by Daniel Chirot and Martin Seligman, 343–362. Washington, D.C.: American Psychological Association.

McCrae, Robert R. 2000. "Trait Psychology and the Revival of Personality and Culture Studies." *American Behavioral Scientist* 44 (1): 10–31.

McGraw, Kathleen M. 2000. "Contributions of the Cognitive Approach to Political Psychology." *Political Psychology* 21 (4): 805–832.

McGraw, Kathleen M., and Thomas M. Dolan. 2007. "Personifying the State: Consequences for Attitude Formation." *Political Psychology* 28 (3): 299–327.

Merari, Ariel, Ilan Diamant, Arie Bibi, Yoav Broshi, and Giora Zakin. 2010a. "Personality Characteristics of 'Self Martyrs'/'Suicide Bombers' and Organizers of Suicide Attacks." *Terrorism and Political Violence* 22 (1): 87–101.

Merari, Ariel, Jonathan Figel, Boaz Ganor, Ephraim Lavie, Yohanan Tzoreff, and Arie Livne. 2010b. "Making Palestinian 'Martyrdom Operations'/'Suicide Attacks': Interviews with Would-Be Perpetrators and Organizers." *Terrorism and Political Violence* 22 (1): 102–119.

Metcalf, Barbara. 1978. "The Madrasa at Deoband: A Model for Religious Education in Modern India." *Modern Asian Studies* 12 (1): 111–134.

Mishal, Shaul, and Maoz Rosenthal. 2005. "Al Qaeda as a Dune Organization: Toward a Typology of Islamic Terrorist Organizations." *Studies in Conflict and Terrorism* 28 (4): 275–293.

Mishali-Ram, Meirav. 2011. "When Ethnicity and Religion Meet: Kinship Ties and Cross-Border Dynamics in the Afghan-Pakistani Conflict Zone." *Nationalism and Ethnic Politics* 17 (3): 257–275.

"Misr Ko Shām Jaisē Hālāt Men Dakhalnē Kī Sāzishen Aur Misrī Āwām Kī Zimmedārīyān" [Conspiracies to Interfere in Egypt Like the Circumstances in Syria and the Responsibilities of the Egyptian People] (Urdu). 2013. *Sharī'at* 2 (17): 2.

Moghadam, Assaf. 2003. "Palestinian Suicide Terrorism in the Second Intifada: Motivations and Organizational Aspects." *Studies in Conflict and Terrorism* 26 (1): 65–92.

Momand, Hassan. 2012. "Barmā Men Mazlūm Musalmānon Ka Qatal-e 'ām Aur Dunyā Kī Pur Asrār Khāmoshī" [The Manslaughter of Oppressed Muslims in Burma and the World's Mysterious Silence] (Urdu). *Sharī'at* 1 (7): 22–27.

——. 2013a. "Mālī Par Maghribī Yalghār: Asbāb o Wujūhāt" [The Western Attack on Mali: Causes and Dimensions] (Urdu). *Sharī'at* 2 (12): 9–11.

——. 2013b. "Turkī Se Muāfī Māngnā . . . Israīl Kā Gharūr Zamīn Būs" [Asking Forgiveness from Turkey . . . Israel's Pride Kisses the Ground] (Urdu). *Sharī'at* 2 (14): 18–19.

——. 2014a. “Daulat-e Islāmīya Kē Khilāf Nayā Faujī Ittihād Aur Us Jang Kā Anjām” [The New Alliance Against the Islamic State and the Conclusion of That War] (Urdu). *Sharī'at* 3 (33): 7–8.

——. 2014b. “Tayyab Erdogān Kē Khilāf Muzāhirē, Kis Kī Sāzish Hai?” [Whose Conspiracy Are the Demonstrations Against Tayyab Erdogan?] (Urdu). *Sharī'at* 3 (25): 21–23.

Monroe, Kristen Renwick, and Lina Haddad Kreidie. 1997. “The Perspective of Islamic Fundamentalists and the Limits of Rational Choice Theory.” *Political Psychology* 18 (1): 19–43.

Moosa, Muhammad. 1998. “What Is the Shara'ee [Legal] Status of the Taliban?” www.taliban.com. Available at: https://web.archive.org/web/19980623171030/http://www.taliban.com/fatwa.htm, accessed December 20, 2013.

Moscovici, Serge. 1973. Foreword to *Health and Illness: A Social Psychological Analysis*, ix–xiv. Edited by Claudine Herzlich. London: Academic Press.

——. 1988. “Notes towards a Description of Social Representations.” *European Journal of Social Psychology* 18 (3): 211–250.

——. 1994. “Social Representations and Pragmatic Communication.” *Social Science Information* 33 (2): 163–177.

——. 1998. “The History and Actuality of Social Representations.” In *The Psychology of the Social*, edited by Uwe Flick, 209–247. Cambridge: Cambridge University Press.

Moscovici, Serge, and Ivana Marková. 1998. “Presenting Social Presentations: A Conversation.” *Culture and Psychology* 4 (3): 371–410.

Mozes, Tomer, and Gabriel Weimann. 2010. “The E-Marketing Strategy of Hamas.” *Studies in Conflict and Terrorism* 33 (3): 211–225.

Muhajir, Abu Salamah Al-. 2013. “To the Jihadis in the West.” *Azan: A Call to Jihad* 1 (4): 25–34.

——. 2014. “Britain on Course for Catastrophe.” *Azan: A Call to Jihad* 1 (5): 16–17.

Muhammad, Maulana Fazhl. 1998. “Who Are Taliban?” www.taliban.com. Available at: https://web.archive.org/web/19980623170948/http://www.taliban.com/whotalib.htm, accessed December 20, 2013.

Muhammadi, Maulana Abdullah. 2013. “Nationalism and Islam.” *Azan: A Call to Jihad* 1 (1): 68–73.

Mujahid, Abdul Hadi. 2014. “Ishgāl-e Amrīkāī Wa Jang-e Qawānīn Dar Afghānistān” [The Occupation of the Americans and the War of Laws in Afghanistan] (Dari). *Haqīqat* 1 (1): 23–25.

Mujahid, Habib. 2015. “AIDS: Amrīkī Jārihīyat Kā Tohfa” [AIDS: A Gift of American Bloodshed] (Urdu). *Sharī'at* 3 (35): 7.

Mukhtar, Ahmad. 2006. “Al-'amalīyāt Al-Istishhādīya Fī Afghānistān Wa Athr-hā Fī Al-Istirātījīya Al-Harbīya Al-Mu'āsira” [Martyrdom Operations in Afghanistan and Their Effect on the Contemporary Military Strategy] (Arabic). *Al-Somood* 1 (1): 11–12.

——. 2008. “Al-Sumūd Tabdā 'āma-Hā Ath-Thālith” [Al-Somood Starts Its Third Year] (Arabic). *Al-Somood* 3 (25): 1–3.

Munir, Haafidh. 2009. “Majla *Al-Somood* Al-Islāmīya Tadkhul 'āma-hā 4” [The Islamic Periodical *Al-Somood* Enters Its Fourth Year] (Arabic). *Al-Somood* 4 (37): 2–3.

Musharraf, Pervez. 2006. *In the Line of Fire: A Memoir*. New York: Free Press.

Naby, Eden. 1988. “Islam within the Afghan Resistance.” *Third World Quarterly* 10 (2): 787–805.

“Nāīn Alavan Kī Gyārahvīn Barsī Kī Bābat Imārat-e Islāmīya Kā Elāmīya” [The Announcement of the Islamic Emirate on the Eleventh Anniversary of 9/11] (Urdu). 2012. *Sharī'at* 1 (7): 6.

Nathan, Joanna. 2009. “Reading the Taliban.” In *Decoding the New Taliban: Insights from the Afghan Field*, edited by Antonio Giustozzi, 27–42. New York: Columbia University Press.

Nawid, Mahmood Ahmad. 2014. “Āh Gaza, Āh Falestīn” [Oh Gaza, Oh Palestine!] (Dari). *Haqīqat* 1 (4): 9–12.

Nawid, Senzil. 1997. “The State, the Clergy, and British Imperial Policy in Afghanistan During the 19th and Early 20th Centuries.” *International Journal of Middle Eastern Studies* 29 (4): 581–605.

——. 2012. “Language Policy and Afghanistan: Linguistic Diversity and National Unity.” In *Language Policy and Language Conflict in Afghanistan and Its Neighbors: The Changing Politics of Language Choice*, edited by Harold F. Schiffman, 31–52. Leiden, Netherlands: Brill.

Nesbitt-Larking, Paul, and Catarina Kinnvall. 2012. “The Discursive Frames of Political Psychology.” *Political Psychology* 33 (1): 45–59.

Nida, Eugene. 2000. “Principles of Translation.” In *The Translation Studies Reader*, edited by Lawrence Venuti, 126–140. New York: Routledge.

Nimrozi, Sa'ad. 2011. “Wa Lammā Qataltu Kāfiran Fī Sabīl Allah Wajadtu Hadhihi Lazza!!!” [And When I Killed an Infidel on the Path of God, I Found This Delight!!!] (Arabic). *Al-Somood* 6 (64): 13.

9/11 Review Commission. 2015. *The FBI: Protecting the Homeland in the 21st Century.*

Nojumi, Neamotollah. 2008. "The Rise and Fall of the Taliban." In *The Taliban and the Crisis of Afghanistan*, edited by Robert D. Crews and Amin Tarzi, 90–117. Cambridge: Harvard University Press.

Nordland, Rod, and Jawad Sukhanyar. 2013. "Taliban Kill Election Official, Then Brag on Twitter." *New York Times*, September 18, p. A13.

Nunan, Patricia. 2002. "A Visit to Mohammed Omar's Former House—2002–02–04." *Voice of America*. Available at: http://www.voanews.com/content/a-13-a-2002–02–04–12-a-66276082/540154.html, accessed December 17, 2013.

Nuri, Aimal. 2013. "Turkī Injēnīr Kī Āzādī Aur Qarāīn o Qīāsāt" [The Freedom of the Turkish Engineer, Contexts, and Analogies] (Urdu). *Sharī'at* 2 (14): 29.

Oakes, Penelope. 2002. "Psychological Groups and Political Psychology: A Response to Huddy's 'Critical Examination of Social Identity Theory.'" *Political Psychology* 23 (4): 809–824.

Oakes, Penelope J., John C. Turner, and S. Alexander Haslam. 1991. "Perceiving People as Group Members: The Role of Fit in the Salience of Social Categorizations." *British Journal of Social Psychology* 30 (2): 125–144.

Obama, Barack. 2013. "President Barack Obama's State of the Union Address—as Prepared for Delivery." Whitehouse.gov. Available at: http://www.whitehouse.gov/the-press-office/2013/02/12/president-barack-obamas-state-union-address, accessed April 11, 2014.

——. 2015. "Remarks by the President in Closing of the Summit on Countering Violent Extremism." White House Office of the Press Secretary. Available at: https://www.whitehouse.gov/the-press-office/2015/02/18/remarks-president-closing-summit-countering-violent-extremism, accessed March 25, 2015.

Obeyesekere, Gananath. 1990. *The Work of Culture: Symbolic Transformation in Psychoanalysis and Anthropology*. Chicago: Chicago University Press.

Omrani, Bijan. 2009. "The Durand Line: History and Problems of the Afghanistan-Pakistan Border." *Asian Affairs* 40 (2): 177–195.

Omrani, Bijan, and Frank Ledwidge. 2009. "Rethinking the Durand Line: The Legality of the Afghan-Pakistani Frontier." *RUSI Journal* 154(5): 48–56.

Onishi, Norimitsu. 2001. "The Missing Mullah." *New York Times*, December 18.

Onwuegbuzie, Anthony J., and Kathleen M. T. Collins. 2007. "A Typology of Mixed Methods Sampling Designs in Social Science Research." *Qualitative Report* 12 (2): 281–316.

Orsini, Francesca. 2012. "How to Do Multilingual Literary History? Lessons from Fifteenth- and Sixteenth-Century North India." *Indian Economic and Social History Review* 49 (2): 225–246.

Payne, Kenneth. 2009. "Winning the Battle of Ideas: Propaganda, Ideology, and Terror." *Studies in Conflict and Terrorism* 32 (2): 109–128.

Perrin, Andrew J. 2005. "National Threat and Political Culture: Authoritarianism, Antiauthoritarianism, and the September 11 Attacks." *Political Psychology* 26 (2): 167–194.

Peters, Gretchen. 2009. *Seeds of Terror: How Heroin Is Bankrolling the Taliban and al Qaeda*. New York: St. Martin's Press.

Pollock, Sheldon. 1995. "Literary History, Indian History, World History." *Social Scientist* 23 (10/12): 112–142.

——. 1998. "The Cosmopolitan Vernacular." *Journal of Asian Studies* 57 (1): 6–37.

——. 2000. "Cosmopolitan and Vernacular in History." *Public Culture* 12 (3): 591–625.

——. 2006. *The Language of the Gods in the World of Men: Sanskrit, Culture, and Power in Premodern India*. Berkeley: University of California Press.

Poole, Fitz John Porter. 1986. "Metaphors and Maps: Towards Comparison in the Anthropology of Religion." *Journal of the American Academy of Religion* 54 (3): 411–457.

Post, Jerrold M. 1979. "Personality Profiles in Support of the Camp David Summit." *Studies in Intelligence* 23: 1–5.

——. 1986. "Narcissism and the Charismatic Leader-Follower Relationship." *Political Psychology* 7 (4): 675–688.

——. 1991. "Saddam Hussein of Iraq: A Political Psychology Profile." *Political Psychology* 12 (2): 279–289.

——. 2003a. "Assessing Leaders at a Distance: The Political Personality Profile." In *The Psychological Assessment of Political Leaders: With Profiles of Saddam Hussein and Bill Clinton*, edited by Jerrold M. Post, 69–104. Ann Arbor: University of Michigan Press.

——. 2003b. "Leader Personality Assessments in Support of Government Policy." In *The Psychological Assessment of Political Leaders: With Profiles of Saddam Hussein and Bill Clinton*, edited by Jerrold M. Post, 39–61. Ann Arbor: University of Michigan Press.

——. 2005. "When Hatred Is Bred in the Bone: Psycho-Cultural Foundations of Contemporary Terrorism." *Political Psychology* 26 (4): 615–636.

——. 2007. *The Mind of the Terrorist: The Psychology of Terrorism from the IRA to al-Qaeda*. New York: Palgrave Macmillan.

——. 2009. "Reframing of Martyrdom and Jihad and the Socialization of Suicide Terrorists." *Political Psychology* 30 (3): 381–385.

Post, Jerrold M., Ehud Sprinzak, and Laurita M. Denny. 2003. "The Terrorists in Their Own Words: Interviews with 35 Incarcerated Middle Eastern Terrorists." *Terrorism and Political Violence* 15 (1): 171–184.

Post, Jerrold M., Stephen G. Walker, and David G. Winter. 2003. "Profiling Political Leaders: An Introduction." In *The Psychological Assessment of Political Leaders: With Profiles of Saddam Hussein and Bill Clinton*, edited by Jerrold Post, 1–7. Ann Arbor: University of Michigan Press.

Potter, Jonathan, and Derek Edwards. 1999. "Social Representations and Discursive Psychology: From Cognition to Action." *Culture and Psychology* 5 (4): 447–458.

"Prison Break: An Exclusive Interview with Adnan Rasheed." 2013. *Azan: A Call to Jihad* 1 (1): 42–49.

Pritchett, F. W., and K. A. Khaliq. 2003. *A Practical Handbook of Urdu Meter.* Madison: University of Wisconsin.

"Pūhantūn-e Amrīkāī-e Afghānistān wa Negrānī Dar Maurid-e Marhala Pas Az Khurūj-e Nīrūhaī-e Khārijī" [The American University of Afghanistan and Concern About the Time of the Phase After the Departure of Foreign Troops] (Dari). 2013. *Srak* 8 (77): 64.

Pye, Lucian W. 1997. "Introduction: The Elusive Concept of Culture and the Vivid Reality of Personality." *Political Psychology* 18 (2): 241–254.

"Qāid Al-Mujāhidīn Bi-Muhāfadha Urūzjān Fī Laqā Ma'a *Al-Sumūd*" [The Leader of the Mujahideen in Uruzgan Province in a Meeting with *Al-Somood*] (Arabic). 2006. *Al-Somood* 1 (3): 20–22.

"Qānūn-e Mana' Hejāb Dar Gharb wa Ta'āruz-e Ashkār-e An Bā Usūl-e Demūkrāsī" [The Law Banning Veils in the West and Its Open Opposition with Democratic Principles] (Dari). 2011. *Srak* 6 (62): 76–77.

Qasim, Muhammad. 2013a. "Destroying the 'Country' Idol: Part 1." *Azan: A Call to Jihad* 1 (3): 23–34.

——. 2013b. "Destroying the 'Country' Idol: Part 2." *Azan: A Call to Jihad* 1 (4): 18–22.

——. 2013c. "On the Road to Khilafah." *Azan: A Call to Jihad* 1 (1): 10–17.

——. 2013d. "Secular Education." *Azan: A Call to Jihad* 1 (2): 37–41.

Qazi, Shehzad H. 2011. "Rebels of the Frontier: Origins, Organization, and Recruitment of the Pakistani Taliban." *Small Wars and Insurgencies* 22 (4): 574–602.

Rabinow, Paul, and William M. Sullivan. 1979. *Interpretive Social Science*. Berkeley: University of California Press.

Rahimi. 2012. "Amrīkī Fauj Men Pāgalpan Aur Khudkushī Kē Barthē Huē Wāqiāt" [The Events of Increasing Insanity and Suicide in the American Military] (Urdu). *Sharī'at* 1 (7): 34.

Rahman, Abdul. 2013. "Yahūdīyon Kā Khilonā" [The Toy of the Jews] (Urdu). *Sharī'at* 2 (17): 36–37.

Rahman, Tariq. 2006. "Urdu as an Islamic Language." *Annual of Urdu Studies* 21: 101–119.

Rasanayagam, Angelo. 2003. *Afghanistan: A Modern History*. New York: I.B. Taurus.

——. 2007. *Afghanistan: A Modern History*. New York: I.B. Taurus.

Rasheed, Adnan. 2014a. "Is Pakistan Going to Be Sandwiched?" *Azan: A Call to Jihad* 1 (6): 22–23.

——. 2014b. "The Mir Ali Massacre: The True Face of the Pakistani Army." *Azan: A Call to Jihad* 1 (5): 26–27.

Rāshid, Abu Sa'id. 2010. "La Tansi Yā Ardha Al-Afghān!" [Don't Forget, Oh Land of the Afghans!] (Arabic). *Al-Somood* 5 (51): 26–27.

Rashid, Ahmed. 1999. "The Taliban: Exporting Extremism." *Foreign Affairs* 78 (6): 22–35.

——. 2000. *Taliban: Islam, Oil, and the New Great Game in Central Asia*. New York: I.B. Taurus.

——. 2001a. "Afghanistan Resistance Leader Feared Dead in Blast." *Telegraph* (London), September 11. Available at: http://www.telegraph.co.uk/news/worldnews/asia/afghanistan/1340244/Afghanistan-resistance-leader-feared-dead-in-blast.html.

——. 2001b. "Intelligence Team Defied Musharraf to Help Taliban." *Telegraph* (London), October 10. Available at: http://www.telegraph.co.uk/news/worldnews/asia/pakistan/1359051/Intelligence-team-defied-Musharraf-to-help-Taliban.html.

——. 2002. *Taliban: Islam, Oil, and the New Great Game in Central Asia*. New York: I.B. Taurus.

——. 2010. "A Deal with the Taliban?" *New York Review of Books*, February 25.

Rasuly-Paleczek, Gabriele. 2001. "The Struggle for the Afghan State: Centralization, Nationalism and Their Discontents." In *Identity Politics in Central Asia and the Muslim World: Nationalism, Ethnicity, and Labor in the Twentieth Century*, edited by Willem van Schendel and Erik J. Zürcher, 149–188. New York: I.B. Taurus.

Ratner, Carl. 1999. "Three Approaches to Cultural Psychology." *Cultural Dynamics* 11 (1): 7–31.

——. 2008. "Cultural Psychology and Qualitative Methodology: Scientific and Political Considerations." *Culture and Psychology* 14 (3): 259–288.

"Rawalpindi Massacre." 2014. *Azan: A Call to Jihad* 1 (5): 29.

Reicher, Stephen. 2004. "The Context of Social Identity: Domination, Resistance, and Change." *Political Psychology* 25 (6): 921–945.

Renshon, Jonathan, and Stanley A. Renshon. 2008. "The Theory and Practice of Foreign Policy Decision Making." *Political Psychology* 29 (4): 509–536.

Renshon, Stanley. 2002. "Lost in Plain Sight: The Cultural Foundations of Political Psychology." In *Political Psychology*, edited by Kristen Renwick Monroe, 121–1840. Mahwah, N.J.: Lawrence Erlbaum.

Renshon, Stanley, and John Duckitt. 1997. "Cultural and Cross-Cultural Political Psychology: Toward the Development of a New Subfield." *Political Psychology* 18 (2): 233–240.

Ricci, Ronit. 2012. "Citing as a Site: Translation and Circulation in Muslim South and Southeast Asia." *Modern Asian Studies* 46 (2): 331–353.

Rid, Thomas. 2010. "Cracks in the Jihad." *Wilson Quarterly*, Winter, pp. 40–48.

Risse, Thomas. 2002. "Transnational Actors and World Politics." In *Handbook of International Relations*, edited by Walter Carlsnaes, Thomas Risse, and Beth Simmons, 255–274. London: Sage.

Roberts, Adam. 2009. "Doctrine and Reality in Afghanistan." *Survival: Global Politics and Strategy* 51 (1): 29–60.

Robinson, Francis. 1997. "Religious Change and the Self in Muslim South Asia Since 1800." *South Asia: Journal of South Asian Studies* 20 (1): 1–15.

Roccas, Sonia, and Marilynn B. Brewer. 2002. "Social Identity Complexity." *Personality and Social Psychology Review* 6 (2): 88–106.

Rosati, Jerel A. 2000. "The Power of Human Cognition in the Study of World Politics." *International Studies Review* 2 (3): 45–75.

"Roshan Mustaqbal Kī Umīd Par" [In Hopes of a Luminous Future] (Urdu). 2014. *Sharīʿat* 3 (24): 15.

Ross, Marc Howard. 1995. "Psychocultural Interpretation Theory and Peacemaking in Ethnic Conflicts." *Political Psychology* 16 (3): 523–544.

——. 1997. "The Relevance of Culture for the Study of Political Psychology and Ethnic Conflict." *Political Psychology* 18 (2): 299–326.

——. 2001. "Psychocultural Interpretations and Dramas: Identity Dynamics in Ethnic Conflict." *Political Psychology* 22 (1): 157–178.

Roy, Olivier. 1990. *Islam and Resistance in Afghanistan*. Cambridge: Cambridge University Press.

——. 2001. "Has Islamism a Future in Afghanistan?" In *Fundamentalism Reborn? Afghanistan and the Taliban*, edited by William Maley, 199–211. New York: New York University Press.

——. 2004. *Globalized Islam: The Search for a New Ummah*. New York: Columbia University Press.

——. 2010. "Headquarters of Terrorists is 'Virtual Ummah.'" *New Perspectives Quarterly* 27 (2): 42–44.

Rubin, Alissa J. 2010. "Taliban Overhauls Image to Win Allies." *New York Times*, January 21, p. A1.

Rubin, Barnett. 2002. *The Fragmentation of Afghanistan: State Formation and Collapse in the International System*. New Haven: Yale University Press.

Rubin, Elizabeth. 2006. "In the Land of the Taliban." *New York Times*, October 22. Available at: http://www.nytimes.com/2006/10/22/magazine/22afghanistan.html?pagewanted=all&_r=0.

Rukhshani, Maulawi Raza. 2013. "Mehmān-Nawāzī Dar Farhang-e Islāmī" [Hospitality in Islamic Culture] (Dari). *Srak* 8 (74): 80–84.

Ruttig, Thomas. 2009. "Loya Pakita's Insurgency." In *Decoding the New Taliban: Insights from the Afghan Field*, edited by Antonio Giustozzi, 57–88. New York: Columbia University Press.

Ryan, Michael W. S. 2013. *Decoding Al-Qaeda's Strategy: The Deep Battle Against America*. New York: Columbia University Press.

Saafi, Zubair. 2008. "Al-Taghayyur Al-Ijtimā'ī Li'l Bī'a Al-Afghānīya Wa Taghrībi-hā" [Social Change for the Afghan Environment and Its Westernization] (Arabic). *Al-Somood* 3 (28): 34–37.

——. 2009. "Al-Mukhatatāt Al-Ijrāmīya Li-Taghayyur Al-Thaqāfa Al-Afghānīya" [The Criminal Plans to Change Afghan Culture] (Arabic). *Al-Somood* 4 (38): 47–48.

Saeed, Abdullah. 2005. *Interpreting the Qur'an: Towards a Contemporary Approach*. New York: Routledge.

Sageman, Marc. 2004. *Understanding Terror Networks*. Philadelphia: University of Pennsylvania Press.

Saigol, Rubina. 2012. "The Multiple Self: Interfaces Between Pashtun Nationalism and Religious Conflict on the Frontier." *South Asian History and Culture* 3 (2): 197–214.

Saikal, Amin. 2012. *Modern Afghanistan: A History of Struggle and Survival*. New York: I.B. Taurus.

Salahshur, Shahid. 2014. "Mānīn-e Jihād Che Mīgūyand?" [What Do the Deniers of Jihad Say?] (Dari). *Haqīqat* 1 (2): 41–42.

Salahuddin, Usama. 2013. Mali. *Azan: A Call to Jihad* 1 (1): 57–59.

Samangani, Inamullah Habibi. 2014. "Majbūrīyat-e Dīmūkrāsī Wa Alamīya-e Musulmānān-e Burmā" [The Compulsions of Democracy and the Affliction of the Muslims of Burma] (Dari). *Haqīqat* 1 (4): 25–26.

Sande, Hans. 1992. "Palestinian Martyr Widowhood—Emotional Needs in Conflict with Role Expectations?" *Social Science and Medicine* 34 (6): 709–717.

"Sar Khordegi Wa Tars-e Hampaimānān-e Amrīkā az Jang-e Afghānistān" [Frustration and Fear of America's Allies from the Afghanistan War] (Dari). 2010. *Srak* 5 (60): 37–38.

"Saudi Arabia Severs Ties with Taliban." 2001. *Guardian*, September 25. Available at: http://www.theguardian.com/world/2001/sep/25/afghanistan.terrorism7.

Savage, Charlie. 2013a. "N.S.A. Said to Have Paid E-Mail Providers Millions to Cover Costs from Court Ruling." *New York Times*, August 23, p. A14.

——. 2013b. "N.S.A. Said to Search Content of Messages to and from U.S." *New York Times*, August 8, p. A1.

Schatz, Robert T, and Howard Lavine. 2007. "Waving the Flag: National Symbolism, Social Identity, and Political Engagement." *Political Psychology* 28 (3): 329–355.

Schiffman, Harold F., and Brian Spooner. 2012. "Afghan Languages in a Larger Context of Central and South Asia." In *Language Policy and Language Conflict in Afghanistan and Its Neighbors: The Changing Politics of Language Choice*, edited by Harold F. Schiffman, 1–28. Leiden, Netherlands: Brill.

Schimmel, Annemarie. 1992. *Islam: An Introduction*. Albany: State University of New York Press.

Schmitt, Eric. 2009. "Many Sources Feed Taliban's War Chest." *New York Times*, October 19.

Schofield, Julian. 2011. "Diversionary Wars: Pashtun Unrest and the Sources of the Pakistan-Afghan Confrontation." *Canadian Foreign Policy Journal* 17 (1): 38–49.

Seibert, Sam. 2012. "That Way Madness Lies; Has Mullah Omar Lost His Mind?" *Newsweek International Edition*, October 22, p. 1.

Shah, Niaz A. 2012. "The Islamic Emirate of Afghanistan: A *Layeha* [Rules and Regulations] for *Mujahidin*." *Studies in Conflict and Terrorism* 35 (6): 456–470.

Shahrani, Nazif M. 2002. "War, Factionalism, and the State in Afghanistan." *American Anthropologist* 104 (3): 715–722.

Shahzad, Syed Saleem. 2010. "War and Peace: A Taliban View." *Asia Times*, March 26. Available at: http://www.atimes.com/atimes/South_Asia/LC26Df03.html.

"Shaikh al-Hadith wa al-Qurān Maulānā Qutaiba (Khāksār) Dar Musāhaba Bā *Srak*" (Sheik of Hadith and Quran Maulana Qutaiba "Khaksar" in an Interview with *Srak*) (Dari). 2012. *Srak* 7 (71): 14–25.

Shamāsina, Alā. 2010. "Al-Sīyāsa Al-Amrīkīa Wa Al-Wujūh Al-Musta'āmara" [American Policy and Colonial Dimensions] (Arabic). *Al-Somood* 5 (54): 51.

Shapiro, Jacob N., and David A. Siegel. 2012. "Moral Hazard, Discipline, and the Management of Terrorist Organizations." *World Politics* 64 (1): 39–78.

"*Sharī'at* Magazīn Ke Aghāz Par Imārāt-e Islāmīya-e Afghānistān Ke Mīdīa Wa Siqāfatī Kamīshan Kā Paighām" [A Message from the Media and Cultural Commission of the Islamic Emirate of Afghanistan on the Start of *Shar'iat* Magazine] (Urdu). 2012. *Sharī'at* 1 (1): 5–6.

Shehzad, Mohammad. 2004. "The Rediff Interview/Mullah Omar." Rediff.com. Available at: http://www.rediff.com/news/2004/apr/12inter.htm, accessed December 17, 2013.

Sher, Muhammad Ali. 2014. "Our Jihadi Heritage." *Azan: A Call to Jihad* 1 (5): 34–35.

Shetret, Liat. 2011. *Use of the Internet for Counter-Terrorist Purposes*. Center on Global Counterterrorism Cooperation. Available at: www.globalct.org, accessed June 7, 2015.

Shweder, Richard A. 1991. *Thinking Through Culture: Expeditions in Cultural Psychology*. Cambridge: Harvard University Press.

——. 1997. "The Surprise of Ethnography." *Ethos* 25 (2): 152–163.

——. 1999. "Why Cultural Psychology?" *Ethos* 27 (1): 62–73.

Shweder, Richard A., and Maria Sullivan. 1993. "Cultural Psychology: Who Needs It?" *Annual Review of Psychology* 44: 497–523.

Sieff, Kevin. 2012. "A Fight for Afghanistan's Most Famous Artifact." *Washington Post*, December 29.

Silke, Andrew. 2003. "The Psychology of Suicidal Terrorism." In *Terrorists, Victims, and Society: Psychological Perspectives on Terrorism and Its Consequences,* edited by Andrew Silke, 93–108. Chichester, U.K.: John Wiley & Sons.

Silverstein, Michael. 1979. "Language Structure and Linguistic Ideology." In *The Elements: A Parasession on Linguistic Units and Levels,* edited by Paul R. Clyne, William F. Hanks, and Carol F. Hofbauer, 193–247. Chicago: Chicago Linguistic Society.

Sinha, Chris. 2000. "Culture, Language, and the Emergence of Subjectivity." *Culture and Psychology* 6 (2): 197–207.

Soherwordi, Syed Hussain Shaheed. 2011. "The Characteristic Traits of Terrorism and Interpretation of Jihad by Al-Qaeda and the Taliban in the Pak-Afghan Pakhtun Society." In *The Dynamics of Change in Conflict Societies: Pakhtun Region in Perspective,* 81–97. University of Peshawar: Hanns Seidel Foundation Islamabad.

South Asian Association for Regional Cooperation. 2014. Home page. Available at: http://www.saarc-sec.org/#, accessed April 29, 2014.

Spears, Russell, and Olivier Klein. 2011. "On Epistemological Shifts and Good Old French Wine (in New Bottles): Commentary on Elcheroth, Doise, and Reicher." *Political Psychology* 32 (5): 769–780.

Speckhard, Anne, and Khapta Akhmedova. 2005. "Talking to Terrorists." *Journal of Psychohistory* 33 (2): 125–156.

——. 2006. "The Making of a Martyr: Chechen Suicide Terrorism." *Studies in Conflict and Terrorism* 29 (5): 429–492.

Spivak, Gayatri Chakravorty. 2000. "The Politics of Translation." In *The Translation Studies Reader,* edited by Lawrence Venuti, 397–416. New York: Routledge.

Staerklé, Christian, Alain Clémence, and Dario Spini. 2011. "Social Representations: A Normative and Dynamic Intergroup Approach." *Political Psychology* 32 (5): 759–768.

Steele, Jonathan. 2011. *Ghosts of Afghanistan: The Haunted Battleground.* Berkeley: Counterpoint.

Stein, Jeff. 2011. "Report: Pakistani Spy Agency Rushed Mullah Omar to Hospital." *Washington Post,* January 18. Available at: http://voices.washingtonpost.com/spy-talk/2011/01/mullah_omar_treated_for_heart.html.

Stern, Paul C. 1995. "Why Do People Sacrifice for Their Nations?" *Political Psychology* 16 (2): 217–235.

Stewart, Devin J. 1990. "Sajʿ in the 'Qurʾān': Prosody and Structure." *Journal of Arabic Literature* 21 (2): 101–139.

Strick van Linschoten, Alex, and Felix Kuehn. 2012a. *An Enemy We Created: The Myth of the Taliban-Al Qaeda Merger in Afghanistan*. New York: Oxford University Press.

——. 2012b. *Poetry of the Taliban*. New York: Columbia University Press.

"Sūba-e Helmand Ke Nā'ib Amīr Mulla Muhammad Dāwūd Muzammil Se Numāinda-e *Sharī'at* Kī Ek Nishast" [A Meeting of a Representative of *Sharī'at* with the Deputy Amīr of Helmand Province Mulla Muhammad Dawud Muzammil] (Urdu). 2011. *Sharī'at* 1 (1): 7–10.

"Sūba-e Helmand Kē Nāib Amīr Mullah Muhammad Dawūd Muzammil Sē Numāinda-e *Sharī'at* Kī Ek Nishast" [A Meeting of the Deputy Governor of Helmand Province Mullah Muhammad Dawud Muzammil with a Representative of *Sharī'at*] (Urdu). 2012. *Sharī'at* 1 (1): 7–10.

Sultana, Aneela. 2009. "Taliban or Terrorist? Some Reflections on Taliban's Ideology." *Politics and Religion* 1 (3): 7–24.

Super User. 2011. "Message of Felicitation of the Esteemed Amir-ul-Momineen on the Occasion of Eid-ul-Odha." Islamic Emirate of Afghanistan. November 4. Available at: http://www.shahamat-english.com/english/index.php/paighamoona/28913-message-of-felicitation-of-the-esteemed-amir-ul-momineen-on-the-occasion-of-eid-ul-odha.

——. 2012. "Message of Felicitation of the Esteemed Amir-ul-Momineen on the Occasion of the Eid-ul-Odha." Islamic Emirate of Afghanistan. October 24. Available at: http://www.shahamat-english.com/english/index.php/paighamoona/28791-message-of-felicitation-of-the-esteemed-amir-ul-momineen-on-the-occasion-of-the-eid-ul-odha.

——. 2013. "How Much Rational Is Obama's Decision About the Stay of American Troops in Afghanistan?" Islamic Emirate of Afghanistan. February 18. Available at: http://shahamat-english.com/english/index.php/articles/29205-how-much-rational-is-obama's-decision-about-the-stay-of-american-troops-in-afghanistan.

Tafshin, Akram. 2013a. "Islām Aur Thīōkrāsī" [Islam and Theocracy] (Urdu). *Shar'īat* 2 (13): 25–26.

——. 2013b. "Māhnāma-e Sharī'at: Āghāz Aur Sāl Kā Safar" [The Monthly *Sharī'at*: The Beginning and Journey over a Year] (Urdu). *Sharī'at* 2 (12): 7.

——. 2014. "Sharī'at Kī Kāmyābī Kā Barā Sabūt Is Kī Barhtī Hūī Maqbūlīyat Hai—Zabīullah Mujāhid" [The Biggest Evidence of Sharī'at's Success Is Its Growing Acceptance—Zabiullah Mujahid] (Urdu). *Sharī'at* 3(24): 17–20.

Tajfel, Henri. 1970. "Experiments in Intergroup Discrimination." *Scientific American* 223 (5): 96–102.

———. 1974. "Social Identity and Group Behavior." *Social Science Information* 13 (2): 65–93.

Tajfel, Henri, and John C. Turner. 1979. "An Integrative Theory of Intergroup Conflict." In *The Social Psychology of Intergroup Relations*, edited by William G. Austin and Stephen Worchel, 33–47. Belmont: Brooks/Cole Publishing Company.

Tajfel, Henri, M. G. Billig, R. P. Bundy, and Claude Flament. 1971. "Social Categorization and Intergroup Behaviour." *European Journal of Social Psychology* 1 (2): 149–178.

"Tala-e Afghānistān" [The Trap of Afghanistan] (Dari). 2010. *Srak* 5 (60): 85–86.

Taliban. 1998. "The Movement of Taliban." www.taliban.com. Available at: https://web.archive.org/web/19980623170905/http://www.taliban.com/talibmov.htm, accessed December 20, 2013.

"Talīm Aur Tarbīat Se Mutalaq Imārat-e Islāmīya Kā Elāmīya" [The Announcement of the Islamic Emirate Related to Education and Upbringing] (Urdu). 2013. *Sharī'at* 1 (11): 21.

Tanwir, Abid. 2014. "Hindūstān, Amrīka Maujūdegī Par Isrār Kyon" [Why Is India Insisting on America's Presence] (Urdu). *Sharī'at* 3 (25): 6.

Tarzi, Amin. 2009. "The Neo-Taliban." In *The Taliban and the Crisis of Afghanistan*, edited by Robert D. Crews and Amin Tarzi, 274–310. Cambridge: Harvard University Press.

Taylor, Max, and John Horgan. 2001. "The Psychological and Behavioural Bases of Islamic Fundamentalism." *Terrorism and Political Violence* 13 (4): 37–71.

Tayyab, Maulana Qari Muhammad. 2013. "Disavowal of the Kuffar: A Quranic Perspective." *Azan: A Call to Jihad* 1 (2): 43–49.

Thackston, W. M. 2008. *The Gulistan of Sa'di*. Bethesda, Md.: Ibex.

Thruelsen, Peter Dahl. 2010. "The Taliban in Southern Afghanistan: A Localised Insurgency with a Local Objective." *Small Wars and Insurgencies* 21 (2): 259–276.

Tomsen, Peter. 2011. *The Wars of Afghanistan: Messianic Terrorism, Tribal Conflicts, and the Failures of Great Powers*. New York: PublicAffairs.

Triandis, Harry C., and Eunkook M. Suh. 2002. "Cultural Influences on Personality." *Annual Review of Psychology* 53: 133–60.

Tsfati, Yariv, and Gabriel Weimann. 2002. "Www.terrorism.com: Terror on the Internet." *Studies in Conflict and Terrorism* 25 (5): 317–332.

Turner, John C. 1975. "Social Comparison and Social Identity: Some Prospects for Intergroup Behavior." *European Journal of Social Psychology* 5 (1): 5–34.

——. 1982. "Towards a Cognitive Redefinition of the Social Group." In *Social Identity and Intergroup Relations*, edited by Henri Tajfel, 15–40. Cambridge: Cambridge University Press.

Turner, John C., and Penelope J. Oakes. 1986. "The Significance of the Social Identity Concept for Social Psychology with Reference to Individualism, Interactionism and Social Influence." *British Journal of Social Psychology* 25 (3): 237–252.

Tyrer, Peter, Roger Mulder, Mike Crawford, Giles Newton-Howes, Erik Simonsen, David Ndetei, Nestor Koldobsky, Andrea Fossati, Joseph Mbatia, and Barbara Barrett. 2010. "Personality Disorder: A New Global Perspective." *World Psychiatry* 9 (1): 56–60.

Umar, Maulana Asim. 2013a. "The Pain of Syria." *Azan: A Call to Jihad* 1 (2): 51–55.

——. 2013b. "To the Muslims of India." *Azan: A Call to Jihad* 1 (2): 41–46.

——. 2014. "Establishing Islam Through Democracy?" *Azan: A Call to Jihad* 1 (5): 19–21.

Urciuoli, Bonnie. 1995. "Language and Borders." *Annual Review of Anthropology* 24: 525–546.

U.S. Consulate (Peshawar). 1994. "New Fighting and New Forces in Kandahar." National Security Archive. November 3. Available at: www2.gwu.edu/~nsarchiv/NSAEBB/NSAEBB97/tal1.pdf.

U.S. Defense Intelligence Agency. 2001. "IIR [Excised] Taliban Leadership, Part 1—Mullah ((Omar)) and the Council of Ministers." November 7. Available at: www2.gwu.edu/~nsarchiv/NSAEBB/NSAEBB295/doc16.pdf.

U.S. Embassy (Islamabad). 1995. "Finally, A Talkative Talib: Origins and Membership of the Religious Students' Movement." National Security Archive. February 20. Available at: www2.gwu.edu/~nsarchiv/NSAEBB/NSAEBB97/tal8.pdf.

——. 1996. "Afghanistan: Taliban Seeks Low Profile Relations with the USG—at Least for Now." National Security Archive. October 8. Available at: www2.gwu.edu/~nsarchiv/NSAEBB/NSAEBB295/doc02.pdf.

——. 1997. "Afghanistan: Taliban Decision-Making and Leadership Structure." National Security Archive. December 30. Available at: www2.gwu.edu/~nsarchiv/NSAEBB/NSAEBB295/doc08.pdf.

——. 1998a. "Afghanistan: Demarche to Taliban on New Bin Laden Threat." National Security Archive. September 14. Available at: www2.gwu.edu/~nsarchiv/NSAEBB/NSAEBB134/Doc%206.pdf.

——. 1998b. "Afghanistan: The Taliban's Decision-Making Process and Leadership Structure." National Security Archive. December 31. Available at: www2.gwu.edu/~nsarchiv/NSAEBB/NSAEBB295/doc13.pdf.

——. 1998c. "Bad News on Pak Afghan Policy: GOP Support for the Taliban Appears to Be Getting Stronger." National Security Archive. July 1. Available at: www.gwu.edu/%7Ensarchiv/NSAEBB/NSAEBB97/talib8.pdf.

——. 2000. "Searching for the Taliban's Hidden Message." National Security Archive. September 19. Available at: www2.gwu.edu/~nsarchiv/NSAEBB/NSAEBB134/Document%206%20-%20ISLAMA%2005749.pdf.

U.S. Intelligence Information Report. 1996. "From [Excised] to DIA Washington D.C. [Excised], Cable [Excised]/Pakistan Interservice Intelligence/Pakistan (PK) Directorate Supplying the Taliban Forces." National Security Archive. Available at: www2.gwu.edu/~nsarchiv/NSAEBB/NSAEBB227/15.pdf, accessed October 17, 2013.

Van Dijk, Teun A. 1993. "Principles of Critical Discourse Analysis." *Discourse and Society* 4 (2): 249–283.

Van Ginneken, Jaap. 1988. "Outline of a Cultural History of Political Psychology." In *The Psychology of Politics*, edited by William F. Stone and Paul E. Schnaffer, 3–22. New York: Free Press.

Victoroff, Jeff. 2005. "The Mind of the Terrorist: A Review and Critique of Psychological Approaches." *Journal of Conflict Resolution* 49 (1): 3–42.

Voice of America. 2001. "Mullah Omar—in His Own Words." *Guardian*, September 26. Available at: http://www.theguardian.com/world/2001/sep/26/afghanistan.features11.

Wahab, Shaista, and Barry Youngerman. 2007. *A Brief History of Afghanistan*. New York: Facts on File, Inc.

Wan, Ching. 2012. "Shared Knowledge Matters: Culture as Intersubjective Representations." *Social and Personality Psychology Compass* 6 (2): 109–125.

Ward, Dana. 2002. "Political Psychology: Origin and Development." In *Political Psychology*, edited by Kristen Renwick Monroe, 61–78. Mahwah, N.J.: Lawrence Erlbaum.

Watson Institute. 2013. "The Costs of 12 Years in Afghanistan." Brown University. Available at: http://watson.brown.edu/news/2013/costs-12-years-afghanistan, accessed October 8, 2013.

Wehr, Hans. 1976. *A Dictionary of Modern Written Arabic*. Edited by J. Milton Cowan. Ithaca, N.Y.: Spoken Language Services.

Weinbaum, Marvin G., and Jonathan B. Harder. 2008. "Pakistan's Afghan Policies and Their Consequences." *Contemporary South Asia* 16 (1): 25–38.

Weiner, Tim. 2001. "Seizing the Prophet's Mantle: Muhammad Omar." *New York Times*, December 7, p. 3.

Weltman, David, and Michael Billig. 2001. "The Political Psychology of Contemporary Anti-Politics: A Discursive Approach to the End-of-Ideology Era." *Political Psychology* 22 (2): 367–382.

White, Jeremy. 2012. "Virtual Indoctrination and the Digihad: The Evolution of Al-Qaeda's Media Strategy." November 19. *Small Wars Journal*. Available at: http://smallwarsjournal.com/jrnl/art/virtual-indoctrination-and-the-digihad.

White House. 2009. A New Strategy for Afghanistan and Pakistan. March 27. Available at: http://www.whitehouse.gov/blog/09/03/27/A-New-Strategy-for-Afghanistan-and-Pakistan.

——. 2015. "Obama, Afghanistan's President Ghani in Joint Press Conference." U.S. Department of State. March 24. Available at: http://iipdigital.usembassy.gov/st/english/texttrans/2015/03/20150324314429.html#axzz3VQjm6CDr.

Whittemore, Robin, Susan K. Chase, and Carol Lynn Mandle. 2001. "Validity in Qualitative Research." *Qualitative Health Research* 11 (4): 522–537.

Williams, Brian Glyn. 2011. "On the Trail of the 'Lions of Islam': Foreign Fighters in Afghanistan and Pakistan, 1980–2010." *Orbis* 55 (2): 216–239.

——. 2012. *Afghanistan Declassified: A Guide to America's Longest War*. Philadelphia: University of Pennsylvania Press.

Winter, David G. 2000. "Power, Sex, and Violence: A Psychological Reconstruction of the 20th Century and an Intellectual Agenda for Political Psychology." *Political Psychology* 21 (2): 383–404.

Wolpert, Stanley. 1982. *Roots of Confrontation in South Asia: Afghanistan, Pakistan, India, and the Superpowers*. New York: Oxford University Press.

Wood, Graeme. 2009. "Security Blanket: Afghanistan's Most Venerable Relic Faces Its Greatest Challenge." *Atlantic*, January 1.

Wright, Lawrence. 2006. *The Looming Tower: Al-Qaeda and the Road to 9/11*. New York: Vintage Books.

Wright, Mathew. 2011. "Diversity and the Imagined Community: Immigrant Diversity and Conceptions of National Identity." *Political Psychology* 32 (5): 837–862.

Yarshater, Ehsan. 2012. "Hafez I: An Overview." *Encyclopaedia Iranica*. Available at: http://www.iranicaonline.org/articles/hafez-I, accessed January 11, 2015.

Yousafzai, Sami, and Ron Moreau. 2004. "Last Days of the Taliban?" *Newsweek*, December 27.

——. 2007. "The Mysterious Mullah Omar." *Newsweek*, March 5.

——. 2010. "The Mullah Omar Show." *Newsweek*, August 16.

Yusufzai, Habibullah. 2013. "Millat-e Mā Hargez Millat-e Ghulām-e Fitrat Nīst" [Our Nation Has Never Been a Nation of a Slavish Nature] (Dari). *Srak* 8 (76): 59.

——. 2014. "Ālām-e Shikast-e Āshkār-e Āmrīkā Dar Jang-e Afghānistān" [The Announcement of the Clear Defeat of America in the Afghanistan War] (Dari). *Srak* 9 (79): 69–70.

Z. N. 2014. "Āmrīkā, Nasl-e Āyanda-e Mā Rā Be Sū-e Tabāhī Mīkashānad" [America Is Dragging Our Future Generation Toward Destruction] (Dari). *Srak* 9 (80): 10–11.

Zaeef, Abdul Salam. 2010. *My Life with the Taliban*. Edited by Alex Strick van Linschoten and Felix Kuehn. London: Hurst.

Zahid-ur-Rashidy, Maulana. 1998. "Osama Bin Ladin, Klashnikov and 'Uncle Sam.'" www.taliban.com. Available at https://web.archive.org/web/19980623170313/http://www.taliban.com/osama.htm, accessed February 6, 2015.

Zakariyya, Muhammad. 2013. "It's a War Against Islam." *Azan: A Call to Jihad* 1 (4): 57–58.

Zaman, Muhammad Qasim. 2002. *The Ulama in Contemporary Islam: Custodians of Change*. Princeton, N.J.: Princeton University Press.

Zaranji, Jamal. 2014. "Būy-e Mushk: Safarnāma-e Jihādī" [The Smell of Perfume: A Jihadist Travelogue] (Dari). *Haqīqat* 1 (2): 25–26.

Zelin, Aaron T. 2014. "The War Between ISIS and al-Qaeda for Supremacy of the Global Jihadist Movement." *Washington Institute for Near East Policy* 20: 1–11.

Zui, Shahid. 2012. "Dīnī Madāris Aur Isrāīlī Skūlz Se Mutalaq Aqwām-e 'Ālam Kā Insāf?" [Religious Schools and the Justice of the World's Nations Related to Israeli Schools?] (Urdu). *Sharī'at* 1 (8): 32–35.

——. 2013a. "Kyā Shām Par Hamlē Kā Khatra Tal Gayā" [Has the Danger of an Attack on Syria Been Postponed?] (Urdu). *Sharī'at* 2 (20): 9–11.

——. 2013b. "Qāhira Kē Musalmānōn Kō Kis Jurm Kī Sazā Dī Gāī" [For What Crimes Is Punishment Being Given to Cairo's Muslims?] (Urdu). *Sharī'at* 2 (19): 11–16.

——. 2014. "Dāish Kā Peshraft Aur Irāq Kā Mustaqbal" [Daish's Advance and Iraq's Future] (Urdu). *Sharī'at* 3 (31): 9–12.

Zurmati, Samiullah. 2014. "Khujasta Bād: Rōz-e Firār-e Ishghālgarān-e Kānādāī Az Afghānistān" [Be Happy! The Day of Retreat of the Occupying Canadians from Afghanistan] (Dari). *Haqīqat* 1 (2): 27–28.

INDEX

Figures indicated by page numbers in italics

Abdali. *See* Ahmad Shah Durrani
Abdullah, Abu, 80–81
Abdullah bin Masood, 75, 154n3
Abdur Rahman Khan (emir), 11, 149n3
Abu Bakr, 57
Abūl Mazaffar Muhi-ud-Dīn Muhammad bin Aurangzeb, 126, 156n1
Adil, Muhammad, 136–137
Afghan, Walid, 19, 123, 126–127
Afghanistan: approach to, xv; areas under Taliban control before 2001, *12*; ethnic groups in, *10*; founding story of, 59; identity politics in, 10, 11, 13; mujahideen resistance movements (1980s), 5, 6, 21; Mullah Omar's priorities for, 51; mullahs in, 59–60; multilingual publications in, 20–21; and Pakistan, 11, 149n3; post-Soviet, 6, 13; social disruptions in, 9; sovereignty of, 19; Taliban rule of, 7–8, 10–11, 13; U.S. involvement in, 6, 13–14, 117
Afghan Shias, 13
Afterlife, 106
Ahmad, Abu, 125
Ahmad, Hafiz Saeed, 94–95
Ahmad, Mirza Ghulam, 86
Ahmad, Mustafa, 67
Ahmad Shah Durrani (king), 59, 126, 156n1
Ahrar, Saiful Adil, 86, 124, 125
AIDS, 81
Akhund, Abdur Rahman, 80
"Alemarah 3" (video), 109–110
Al-Jazeera, 80
Al Qaeda, xiii, xiv, 13–14, 17, 33. *See also* bin Laden, Osama
Al-Rasheed Trust, 9, 149n2
"Al-Sādiqīn" (The Sincere; video), 111–112
Al-Somood (Arabic periodical): casualty lists in, 93–94; eulogy for al-Zarqawi, 152n7; expository text on jihad in, 104–105; on in-group membership, 69–71; and Internet, 26; interview with Haqqani, 71–72, 95–97; martyr biographies in, 98–103; on Muslim rights, 69–70; number of issues analyzed, xv; on out-groups, 84; on Palestine, 128; on protecting Islamic values, 79–80; prototypical group members in, 71–73; purpose of, 26; on Soviet Union defeat, 131; on Turkey, 133. *See also* Arabic texts
Amanullah (king), 11
American University of Afghanistan, 81–82
Amir ul-Momineen, *see* Commander of the Faithful
analogous reasoning, 141
Anderson, Benedict, 18–19, 20, 21
Anushirvan (king), 104
Anwar, Ikrimah, 105–106, 130
Arabic: Taliban use of, 26, 111, 144. See also *Al-Somood* (Arabic periodical); Arabic texts
Arabic literature: rhymed prose (*saja*) in, 70–71
Arabic texts: expository texts on jihad, 108; on in-group membership, 69; jihadist videos, 110–111; martyr biographies, 98, 100, 103; on

Arabic texts (*continued*)
protecting Islamic values, 79–80; on prototypical group members, 71. See also *Al-Somood* (Arabic periodical)
Army of Muslims (Jaish ul-Muslimeen), 49
Asim, Ilyas, 78, 79
Asim, Muhammad, 68–69
Attraction and Absorption Commission, 5
Aurangzeb, Abūl Mazaffar Muhi-ud-Dīn Muhammad bin, 156n1
Azan (English periodical): number of issues analyzed, xv; on prototypical group members, 74–75; on Shia Muslims, 86; on Taliban as global movement, 67–68; on unified moral community, 24–25. *See also* English texts
Azeemullah's biography, 98–100, 103

Badakhshani, Saeed, 84–85
Badakhshi, Nur Allah, 77–78, 79
Badr, Battle of, 80, 99, 156n2
"Badr Brigades 3" ("Katāib Badr 3"; video), 110–111
Baghdadi, Abu Bakr al-, 152n7
Balochi, Saadullah, 106
Bamiyan Buddhas, 45, 50
Baradar, Abdul Ghani, 45–46
Bashar, Haji, 30, 45
Bawadi, Ahmad, 105
Bergdahl, Robert "Bowe," 109–110
Bhabha, Homi, 146
bias, xix, 41
bin Laden, Osama, 13, 32–34, 45, 60, 150n4. *See also* Al Qaeda
biographies, of dead militants, 98–104
Bockstette, Carsten, 92
Boston Marathon bombing, 154n1
Boulding, Kenneth, 118–119
Brahimi, Lakhdar, 48
Buddhas, Bamiyan, 45, 50
Burma, 120–121, 140

Canada, 121–122, 141
casualty lists, 93–95, 115
Center on Global Counterterrorism Cooperation, 145
Chechen insurgents, xiv
Chen, Hsinchun, 108
Christians, 84
cloak, Prophet Muhammad's, 57, 59
cognitive information processing, 118, 141, 142
cognitive psychology, 118, 139–140
Commander of the Faithful (Amir ul-Momineen): al-Baghdadi as, 152n7; history of, 57, 152n7; Mullah Omar as, 32, 40, 44–45, 48, 50, 57–58, 63
conceptual complexity, 56
controlling events, belief in, 55
core values, of Taliban: introduction to, 75–76; on hospitality, 83; on need for Islamic values, 79–83; on shortcomings of Western legal order, 76–79
cosmopolitan languages, 28, 144
counternarratives, to terrorism, 143, 145–146
cross-cultural research, 35–36
cultural analysis, 16, 29, 38, 41, 57
cultural context, 38, 92
cultural psychiatry: contribution of, 147; on cross-cultural measurement, 35, 41; and imagined community theory, 18; introduction to, xv; on militant group discourses, xviii; on personality disorders, 37; on political identities in globalized world, 16–17. *See also* political personality profiles
cultural psychology, 147
culture: definition of, xviii, 16, 66, 146; and personality traits, 35

Dari: in Afghanistan, 11, 13; during Taliban rule, 13, 20; Taliban use of, 26–27, 64, 112, 144. *See also* Dari texts; *Haqīqat* (Dari periodical); *Srak* (Dari-Pashto periodical)
Dari texts: on defending core Islamic values, 81–83; expository texts on jihad in, 107–108; on in-group membership, 68–69; interviews with Taliban officials, 95; jihadist videos, 111–112; martyr biographies in, 103; on out-groups, 84–85; prototypical group

members in, 73–74. See also *Haqīqat* (Dari periodical); *Srak* (Dari-Pashto periodical)
Darul Uloom Haqqani (madrasa), 8
data saturation, xvii
Dawood, Brother, 74
Deoband sect, 8–9, 62, 92, 100, 146
Diagnostic and Statistical Manual of Mental Disorders (*DSM-IV*), 36–37, 42
discourse, xvi, xviii, 144
discourse analysis, xvi, xviii, 29, 146
distrust, of others, 56
Dost Muhammad (emir), 57
Dostum, Abdul Rashid, 13
dream interpretation, 38–39
DSM-IV (*Diagnostic and Statistical Manual of Mental Disorders*), 36–37, 42
Durand Line, 11, 149n3

Educational Commission, 5
egocentric societies, 37
Egypt, 122–124
emirate: definition of, 1
England, 124, 141
English: Taliban use of, 64, 110, 144. See also *Azan* (English periodical); English texts
English texts: expository texts on jihad in, 105–106, 108; on in-group membership, 68; jihadist videos, 109–110; prototypical group members in, 74; readership online, 2. See also *Azan* (English periodical)
ethnic nationalism, 66
expository texts, on jihad, 104–108, 115

Faiz, Ahmed, 69
farmān (orders), 43
Ferguson, James, 22
Fischerkeller, Michael, 119–120, 139–140, 141
followership, 39
Fort Hood shooting, 154n1
Foucault, Michel, xviii
France, 125–126, 129, 141
Freud, Sigmund, 38, 151n3
fund-raising, 64

Gaza, 134
genres, literary: casualty lists, 115; expository texts on jihad, 104–108, 115; interviews with Taliban officials, 95–98, 115; jihadist videos, 108–115; *maghāzī*, 98; martyr biographies, 98–104; poetry, 103–104; rhymed prose (*saja*) in, 70–71; travel, 73
Ghani, Ashraf, 145
Ghaznavi, Shihabuddin, 70, 72, 136
group identity formation, 18–19
group membership, 67–71
Gupta, Akhil, 22

Habib, Qari, 122
Hafez, Khwaja Shamsuddin Muhammad, 103
Hakimzad, Abdul Rahman, 82–83
Hamas, xiv, 128
Hāmid, Mustafa, 121, 137–138
Hammad, Hanif, 136
Hammadi, Badruddin, 128
Haq, Abdul, 8
Haqīqat (Dari periodical): casualty lists in, 94; jihadist travelogues in, 73; number of issues analyzed, xv; purpose of, 26–27. *See also* Dari texts
Haqqani, Badruddin, 104
Haqqani, Jalaluddin, 71–72
Haqqani network, 49, 71
Hashim, Mullah, 46
Hassan, Nidal, 154n1
Health Commission, 4
Hekmatyar, Gulbuddin, 11, 13, 20
Hermann, Margaret, 41–42, 55
Hermann, Richard, 119–120, 139–140, 141
Hikmat, Abdul Rauf, 104
Hindus, 86
Hizb-e Islami, 20, 27
hospitality, 83
Huddy, Leonie, 66–67
Hussain, Jaffer, 67, 135
Hussaini, Maulana Muawiya, 105–106

ibn Abi Talib, Ali, 132, 157n2
Ibn al-Faisal, Turki, 48
ibn al-Khattab, Umar, 57, 58, 132, 157n2
ibn Affan, Uthman, 132, 157n2
ibn Ali, Hussain, 132, 157n2
ibn Ali, Hassan, 132, 157n2
Ibn Qāsim, Muhammad, 126, 156n1
Ibn Qāsim Al-Thaqafī, Imād Al-Dīn Muhammad, 126, 156n1
Ibn Saba, Abdullah, 132, 157n3
Ibn Sebüktegīn,Yamīn Ud-Dawla Abul-Qāsim Mahmūd, 126, 156n1
Ibn Taymiyyah, 25
ideal-hungry follower, 152n4
identity. *See* Taliban identity
identity, Muslim: and Internet, 17–18; and jihad, 105
identity, social: Huddy on assessing, 66–67; and Internet, xiv, 16–17; psychology of, 65–66. *See also* in-groups and out-groups
image theory, 118–120, 139–140, 141, 142
imagined community theory, 18–19, 20
immigrants, for jihad, 74–75
immorality, 7
Imran, Ustadh Abu, 75
India, 126–127, 140
influence and power, need for, 55
in-group bias, 41, 56
in-group favoritism, 141
in-groups and out-groups: approach to, 64–67, 88–89; and core values, 75–83; and group membership, 67–71; and international relations, 117–118; and Mullah Omar, 49, 56, 59; and nationalism, 27, 68–69, 112; and out-groups, 83–87, 117; and prototypical group members, 71–75; and religion, 27, 87, 110, 111; variation of by language, 87. *See also* Taliban international relations
Innocence of Muslims (film), 84
intercession, 100
international relations: and cognitive psychology, 118; and image theory, 118–120, 139–140, 141, 142. *See also* Taliban international relations
Internet: and identity production, xiv, 16–17; and Muslim identity, 17; and Taliban's virtual emirate, 21–23, 29, 143–144; use of by militant groups, 17–18. *See also* virtual emirate
intertextuality, 104
interviews, with Taliban officials, 95–98, 115
Invitation and Guidance Commission, 5
Iqbal, Muhammad, 60, 106
Iran, 13
Iraq, 6
ISIS (Islamic State), 152n7
Islamic Emirate of Afghanistan. *See* Taliban
Islamist resistance groups. *See* mujahideen resistance movements (1980s)
Israel, 51–52, 127, 134, 140. *See also* Jews

Jaish ul-Muslimeen (Army of Muslims), 49
Jalal, Qazi Abdul, 103
Jalaluddin Khwarizm, 154n5
Jamaat Al-Tawhid Wa-l-Jihad, 152n7
Jamiatul Uloomil Islamiyyah (madrasa), 8
Janbaz, Muhammad Farhad, 132–133
Jews, 84–85, 127. *See also* Israel
jihad: approach to in Taliban texts, 91, 115–116; and casualty lists, 93–95; expository texts on, 104–108, 115; and interviews with Taliban officials, 95–98; and jihadist videos, 108–115; long-term goal of, 92; and martyr biographies, 98–104; motivations for, 91–92; studies on Taliban interpretations of, 92; Taliban interpretation of, 90
"Jihad" (video), 113–115
jihadist videos. *See* videos, jihadist
Jihadwal, Maulawi Abdul Hanan, 104

Karzai, Hamid, 14, 47
"Katāib Badr 3" ("Badr Brigades 3"; video), 110–111
Kavkaz Center, 152n7
Khan, Ismail, 13
Khan, Zaheer, 125
Khattak, Khushal Khan, 104
Khel, Painda, 60

Khorasani, Hafiz Muhammad, 86–87
Kinnvall, Catarina, 39, 151n3

Lacan, Jacques, 151n3
language: cosmopolitan and vernacular, 28, 144; and cultural identity, xv, xvi; in framing messages, 144; in-group and out-group identities expressed through, 87; parochial, 28; and social exchange, 88–89; Taliban use of, 64, 88–89, 92. *See also* linguistic ideology; linguistic nationalism
Lanham, Trevor, 40
"laws of the Father" phrase, 39, 151n3
leaders, psychology of. *See* political personality profiles
Liberation Tigers of Tamil Eelam, 1
linguistic ideology, 144
linguistic nationalism, 19–20, 21, 28
literarization, 92
literary culture, xiv
literature. *See* genres, literary; Pashto literature
Ludhyanvi, Rasheed Ahmad, 7–8, 52–53

madrasas, 8–9, 149n1
maghāzī literature, 98, 114
Mahmood Ghaznavi, 126, 156n1
Maiwandi, Akram, 70, 100
Mali, 129
Mansoor, Akhtar, 145
Mansur, Saifur Rahman, 103
Mansur network, 49
martyr biographies, 98–104
martyrdom, 85–86, 95, 105–106, 115. *See also* casualty lists; suicide bombing
Martyrs, Orphans and Crippled Affairs, Commission of, 4–5
Massoud, Ahmed Shah, 13, 54
maximal exclusive paradigm, 88
mental health: and terrorism, xiii
method bias, 41. *See also* bias
methodology: for analysis of Mullah Omar, 42–44; cognitive psychology, 118, 139–140; cosmopolitan, vernacular, and parochial languages, 28; counter-narratives to terrorism, 145; cultural analysis, 16, 29, 38, 41, 57; discourse analysis, xvi, xviii, 29, 146; finding Taliban websites, xiv; image theory, 118–120, 139–140, 141, 142; imagined community theory, 18–19, 20; on language and cultural identity, xv; limiting bias, xix; linguistic nationalism, 19–20, 21, 28; literarization, 92; locating contexts of texts, xiv–xv; motivations for jihad, 91–92; overview of analysis, xviii–xix, 15–17, 147; political psychology, 15–16, 18, 41–42, 66; psycho-biography, 36–37; qualitative research, 42–43; selecting texts for analysis, xvi–xvii; social identity, 65–67; textual content analysis, 41–42; translation of texts, xvii–xviii. *See also* cultural psychiatry; in-groups and out-groups; political personality profiles; virtual emirate
militant groups: and Internet, 1, 17–18; and videos, 108. *See also* Al Qaeda; mujahideen resistance movements (1980s); Taliban
minimal common paradigm, 88
mirror-hungry leader, 152n4
Momand, Hassan, 121, 129, 134, 152n7
morality and virtue, 24–26
Morsi, Mohamed, 123–124
Moscovici, Serge, 88
Muhajir, Abu Salamah Al-, 106
Muhammad (prophet), 57, 58, 59
Muhammad, Maulana Fazhl, 149n1
Muhammad Hashim, 11
Muhammadi, Maulana Abdullah, 25–26, 85–86
Mujahid, Abdul Hadi, 78
Mujahid, Habib, 81
mujahideen, 53, 131
mujahideen resistance movements (1980s), 5, 6, 21
Mukhtar, Ahmad, 26, 104–105
Mukhtar, Umar, 154n5
mullahs, 59–60
Munir, Haafidh, 84
Musharraf, Pervez, 50
Muslim identity. *See* identity, Muslim

Mutanabbi, Al-, 104
Muttalib, Abdul, 95–97
Muttawakil, Ahmed Wakil, 49
Muzammil, Muhammad Dawud, 24, 97–98

Nabi, Ghulam, 100–103
nationalism: Anderson's approach to, 18–19, 20; ethnic, 66; and in-group appeals in Dari texts, 27, 68–69, 112; Mullah Omar on, 52; Muslim critiques of, 25–26, 76–77
nation-state, 19, 120, 140
Nawid, Mahmood Ahmad, 128
Nawruz festival, 86–87
neo-Taliban, 14, 150n5, 151n6
Nesbitt-Larking, Paul, 39, 151n3
Nimrozi, Sa'ad, 72–73
9/11 attacks, 136
9/11 Review Commission, 147
Nixon, Richard, 149n3
Noor-ud-Din Zangi, 154n5
Northern Alliance, 13
Nuri, Aimal, 133–134

Obama, Barack, 117, 135, 138, 143, 145
Omar, Mullah: approach to, xvi–xvii, 35–36, 42–44, 61; analysis and summary of, 55–57, 60–61, 62–63; and Ahmad Shah Durrani, 59; on Bamiyan Buddhas, 45, 50; belief of in controlling events, 55, 58; and bin Laden, 32–33, 45; biographical milestones in leadership of, 44–46; character and drives of, 48–49; as Commander of the Faithful, 32, 40, 44–45, 50, 52–53, 57, 58; conceptual complexity of, 56; cultural analysis of, 57–61; death of, 34; decision-making style of, 54; distrust of others by, 56–57, 59–60; domestic priorities of, 51; dreams of, 38–39, 55, 151n2; and *DSM* categories, 37, 42; emotional reactions of, 47–48; eulogy for al-Zarqawi, 152n7; facts known about, 31; health of, 47; in-group bias of, 56, 59; intellectual capacity of, 47; international views of, 51–52; interpersonal relations of, 45–46, 49; on leadership and power, 50; leadership style of, 52–54; legitimacy of, 30–31; lifestyle of, 46–47; loss of eye, 44; motivation of for office, 56; and Muhammad's cloak, 59; as mullah, 59–60; on nationalism, 52; need of for power and influence, 55; negotiating style of, 54; and opium production, 150n5; orders of, 43, 61, 92–93; personality cult of, 145; personality profile of, 46–49; photograph of, *32*; physical description of, 46; and poetry of Muhammad Iqbal, 60; political outlook of, 50; political personality profile of, 40–41; reputation of, 34; scholarship on, 38; and self-confidence, 55; strategy of towards followers, 53; as Taliban leader, 44; texts from, 43–44; and textual content analysis, 42; theological interpretations by, 58; use of in propaganda, 34–35; US intelligence interest in, 30–31; verbal style of, 53–54; working style of, 46; worldview of, 49–52
Operation Al-Farooq, 58
Operation Khaibar, 154n8
Operation Khalid bin Walid, 58
opium, 150n5
orders (*farmān*), 43
Organization of Islamic Cooperation, 86
out-groups, Taliban on, 83–87. *See also* in-groups and out-groups

Pakistan: and Afghanistan, 11, 149n3; Afghan orphans in, 9; and Qadiani Muslims, 86, 155n6; and Taliban, 7–8, 9, 11, 49, 62, 149n1; Taliban on, 129–131, 140–141
Pakistani Taliban (TTP), 14, 62, 151n6
Palestine, 128, 140
parochial languages, 28
Pashto: in Afghanistan, 11, 13; and Pashtun identity, 10; *sarkozy* in, 125; Taliban use of, 27, 61. See also *Shahāmat* (Pashto periodical); *Srak* (Dari-Pashto periodical)
Pashto literature, 103
Pashtuns, 10–11, 13

periodicals: used in analysis, xv. See also *Al-Somood* (Arabic periodical); *Azan* (English periodical); *Haqīqat* (Dari periodical); *Shahāmat* (Pashto periodical); *Sharī'at* (Urdu periodical); *Srak* (Dari-Pashto periodical)
personality, 35, 36, 38, 41–42, 42–43. *See also* personality disorders; political personality profiles
personality disorders, 36–37, 42
poetry, 103–104
political personality profiles: across cultures, 37–38; development of, 36–37; leadership types within, 40, 152nn4–5; of Mullah Omar, 40–41; schools of thought on, 35–36; strengths of, 42, 46
political psychology, 15–16, 18, 41–42, 66
Pollock, Sheldon, xiv, 28, 92, 93, 144, 145
Post, Jerrold, 36–37, 38, 40, 42, 46. *See also* political personality profiles
power and influence, need for, 55
Prevention of Civilian Casualties, Commission for the, 5
print mediums, 18, 19–20, 21
Prisoners' Affairs, Commission of, 4
prototypical group members, 71–75
psychiatry, 35, 146. *See also* cultural psychiatry
psycho-biography, 36–37
psychology, 35, 37–38, 38–39, 146. *See also* cognitive psychology; cultural psychology; political psychology; social identity theory
Punjabi Taliban, 151n6

Qadiani Muslims, 85–86, 155n6
Qasim, Muhammad, 76–77, 83, 84
qualitative research, 42–43
Qudus, Maulawi Abdul, 103
Quran, 91
Qutaiba, Maulana, 107–108
Qutb, Muhammad, 25

Rabbani, Burhanuddin, 13
Rabbani, Mohammad, 45, 49
Rabi 'ibn 'Amir, 76
Rahimi, 138
Rahman, Abdul Rahman, 85
Rasheed, Adnan, 74, 130, 131
Rāshid, Abu Sa'id, 70–71
religion: and identity, 27, 87, 110, 111, 130, 131
religious scholars, 9, 130–131, 141
reparative leadership, 152n5
rhymed prose (*saja*), 70–71
Rohingya people, 120–121
Rukhshani, Maulawi Raza, 83
Russia, 13, 131–132

Saafi, Zubair, 79–80
Sa'di, 103
saja (rhymed prose), 70–71
Saladin (Salahuddin Ayyubi), 154n5
Salahshur, Shahid, 108
Salahuddin Ayyubi (Saladin), 154n5
Saleem, Moulavi Mohammad, 2
Samangani, Inamullah Habibi, 121
Sarkozy, Nicolas, 125
Saudi Arabia, 49
SCT (self-categorization theory), 65–66, 88
seeking office, motivation for, 56
self, conceptions of, 37
self-categorization theory (SCT), 65–66, 88
self-confidence, 55
Shafeeq, Maulavi Muhammad, 96
Shahāmat (Pashto periodical), 27. *See also* Pashto
Shamāsina, Alā, 134–135
Shamil, Iman, 154n5
Shari'a law, 7–8, 91
Sharī'at (Urdu periodical): approach to, xvi; on 9/11 anniversary, 15; on Burmese Muslims, 120–121; casualty lists in, 94–95; difficulties producing, 21; interview of Helmand province deputy governor, 24, 97–98; on jihad, 107; Mullah Omar's orders in, 43; number of issues analyzed, xv; purpose of, xv, 19, 27; on Taliban as people's champion, 19; theological interpretations by Mullah Omar in, 58; on U.S., 137. *See also* Urdu texts
Sher, Muhammad Ali, 83–84, 154n5
Shia Muslims, 85–86, 157n2
Shias, Afghan, 13

"Sincere, The" (The Al-Sādiqīn), 111–112
SIT (social identity theory), 65–66, 88, 118–119
social exchange, 88–89
social identity. *See* identity, social
social identity complexity, 87
social identity theory (SIT), 65–66, 88, 118
social psychology, 118
social representations theory (SRT), 66, 88
sociocentric societies, 37
Soherwordi, Syed Hussain Shaheed, 92
sovereignty, 18–19, 24
Soviet Union, 13, 131–132
spaces: conceptions of, 22
Srak (Dari-Pashto periodical): cultural xenophobia in, 81–82; on France, 125–126; on jihad, 107–108; masthead of, 21, *22*; number of issues analyzed, xv; on Pakistani religious scholars, 130–131; on U.S., 135–136. *See also* Dari texts; Pashto
SRT (social representations theory), 66, 88
state governance, 24
state nationalism, 18
suicide bombing, xiii, 91, 105–106. *See also* martyrdom
Sultana, Aneela, 92
Sunni Muslims, 157n2
Syria, 132–133

Tafshin, Akram, 21, 27, 77
Tajfel, Henri, 65, 88
Tajikistan, 13
Tajiks, 10, 13
Taliban: approach to, xvi; administrative structure of, 4–5, 14, 19, 60–61; and Afghan identity politics, 10–11; after 9/11, 13–15, 19; and Al Qaeda, 33; and bin Laden, 32–34, 150n4; charitable appeal by, 9; cultural system of, 146–147; and Deobandi sect, 8–9; electronic communications by, 14–15; formation of, 6–7, 9, 31; ideological basis of, 5, 6–7, 92; and imagined community theory, 18–19; initial support for, 7–8; and ISIS, 152n7; leadership of, 8; on mullahs, 60; and nationalism, 25–26, 27; operations naming convention, 154n8; and opium production, 150n5; and Pakistan, 7–8, 9, 11, 49, 149n1; and Pashtuns, 10–11; psychological drives of followers, 39; rule over Afghanistan by, 7–8, *12*, 13; scholarship on, 5–6, 15; self-presentation of, 19, 21, 121, 144; struggle for sovereignty by, 23–24; U.S. intelligence interest in, 30. *See also* neo-Taliban; Omar, Mullah; Pakistani Taliban; Punjabi Taliban; Taliban discourses; Taliban identity; Taliban international relations; Taliban texts; virtual emirate
Taliban discourses: approach to, 15–16, 92–93; approach to identities in, 64–65; counternarratives to, 145–146; ideological flexibility in, 6; and Internet, 22–23, 29, 143–144; and social identities, 65–66; and YouTube, 23. *See also* Taliban identity; Taliban international relations; Taliban texts; virtual emirate
Taliban identity: approach to, 64–65, 66–67; core values of, 75–76; and group membership, 67–71; and Islamic law and international legal order, 76–79; maximal exclusive paradigm of, 88; on out-groups, 83–87; and protecting core Islamic values, 79–83; and prototypical group members, 71–75, 87; and religion, 27, 87, 110, 111, 130, 131; as varied by language, 87. *See also* in-groups and out-groups
Taliban international relations: approach to, 117–118, 120; analysis of, 139, 140–42; on Burma, 120–121, 140; on Canada, 121–22, 141; cognitive processes in, 141; on Egypt, 122–124; on England, 124, 141; on France, 125–126, 129, 141; on India, 126–127, 140; on Israel and Palestine, 127–128, 134, 140; on Mali, 129; overview of, 49; on Pakistan, 129–131, 140–141; on Russia, 131–132; on Syria, 132–33; on Turkey, 133–134, 140; on U.S. as degenerate power, 136–138; on U.S. foreign policy, 134–136, 141; on U.S. Muslims, 138–139

Taliban texts: approach to, xiv, xvi–xvii, 147; analysis of, xviii–xix; avoiding bias in analysis of, xix; biographies of dead militants, 98–104; casualty lists in, 93–95; importance of, xiii–xiv; interviews with Taliban officials, 95–98; jihadist videos, 108–115; locating context of, xiv–xv; multilingual nature of, 20–21, 27–28, 64, 88–89, 92, 144–145; objectives and research questions on, 15, 17; scholarship on, xiv; selecting texts for analysis, xv–xvii; translation of, xvii–xviii; websites, xiii, xiv, 1–2, *3*, 4–5, 11. *See also* genres, literary; Taliban discourses; *specific periodicals*
Tamil Tigers (Liberation Tigers of Tamil Eelam), 1
Tanwir, Abid, 127
Tawheed (oneness), 25
terrorism: 9/11 Review Commission on, 147; counternarratives to, 143, 145–146; and mental health, xiii. *See also* War on Terror
textual content analysis, 41–42
Tipu Sultan, 154n5
translation, xvii–xviii
travel literature, 73
Tsarnaev, Tamerlan and Johar, 154n1
TTP (Tehrīk-e-Talibān-e-Pakistān), 14, 62, 151n6
Turkey, 133–34, 140
Turner, John, 65, 88

Umar, Maulana Asim, 77, 126, 132
United Arab Emirates, 49
United States of America: in Afghanistan, 13–14, 117, 138, 145; and ISIS, 152n7; Mullah Omar on, 51–52, 53–54; and Syria, 133; Taliban on degeneracy of, 136–38; Taliban on foreign policy of, 134–136, 141; Taliban on Muslims in, 138–139; YouTube video of urinating U.S. soldiers, 23
Urdu: as "Islamic language," 61–62; as parochial, 28; as Taliban second language, 13; Taliban use of, 27, 28, 62, 64, 107, 144. See also *Sharīat* (Urdu periodical); Urdu texts
Urdu texts: on defending core Islamic values, 80–81; expository texts on jihad in, 106–107, 108; interviews with Taliban officials, 97; jihadist videos, 113–115; martyr biographies in, 103–104; on outgroups, 85; readership online, 2. See also *Sharīat* (Urdu periodical)
Usman. *See* Ibn Affan, Uthman
Uzbekistan, 13
Uzbeks, 10, 13

Vendrell, Francesc, 16
vernacular languages, 28, 144
videos, jihadist: approach to, 108–109; Arabic, 110–111; Dari, 111–112; English, 109–110; Urdu, 113–115
virtual emirate: approach to, 28–29; for assertion of theological expertise, 116; as global platform, 62; and Internet, 21–23; Mullah Omar's statements on, 62–63; and Taliban's quest for legitimacy, 143–144; and Taliban's struggle for sovereignty, 23–24; and virtue, 24–26
virtue and morality, 24–26

War on Terror: framing of, 88; and image theory, 140; Taliban on, xiii, 77–78, 81. *See also* terrorism
Wehr, Hans, 1

YouTube, 23
Yusufzai, Habibullah, 69, 138

Zaeef, Abdul Salam, 44, 45, 47, 50, 54
Zahid-ur-Rashidy, Maulana, 150n4
Zahir, Abu, 27
Zahir Shah (king), 149n3
Zakariyya, Muhammad, 138–139
Zaranji, Jamal, 73–74
Zarqawi, Abu Musab al-, 152n7
Zoroastrianism, 86–87
Zui, Shahid, 123, 127, 133, 152n7
Zurmati, Samiullah, 121–122